SolidWorks for the Sheet Metal Guy
Course 2: Hole Patterns and Notches

Joe Bucalo
Neil Bucalo

www.SheetMetalGuy.com

SolidWorks for the Sheet Metal Guy
Course 2: Hole Patterns and Notches
Published by
Sheet Metal Guy, LLC
P O Box 498283
Cincinnati, OH 45249
www.SheetMetalGuy.com

Every effort has been made to ensure that all information contained within this book is complete and accurate. However, Sheet Metal Guy assumes no responsibility for the use of said information, nor any infringement of the intellectual property rights of third parties which would result from such use.

Please visit our website at www.SheetMetalGuy.com.

Manufactured in the United States of America.

All technical illustrations and CAD models in this book were produced using SolidWorks 2006. The commands throughout this book have been tested for compatibility with SolidWorks 2007.

ISBN 978-0-9795666-2-2

About the Authors

Joe Bucalo is the founder and President of Applied Production, Inc. He has over 30 years experience in the sheet metal industry, working with and developing software products to change the manufacturing world. When CAD was just beginning on the PC, Joe played a major role in the development of ProFold, the first truly automatic 3D sheet metal unfolding program. When most people were using a wall chart of bend deductions, Joe was promoting the use of the K-factor for more accurate flat patterns.

Joe later paved the way for graphics based sheet metal CAM when he introduced ProFab, which allows the direct transfer of geometry from CAD to CAM. ProFab was the first CAM program in the sheet metal industry to include an automatic tool selection routine.

Joe continues to work with clients to solve their design and manufacturing problems. He has a thorough knowledge of the most popular CAD programs and understands the issues faced by sheet metal manufacturers.

Neil Bucalo has a diverse background, including mechanical design engineering, CAD/CAM support and training, engineering consulting, web development, and technical writing. Neil started his career in support and training of the CADKEY software at Computer Aided Technology, Inc. He then moved forward as a Certified SolidWorks Support Technician.

Upon moving to the Cincinnati area, Neil joined Applied Production, Inc., a SolidWorks Solution Partner, where he has provided customer support and written several user training documents. He also created and served as Editor of the CKD Magazine, dedicated to users of the CADKEY software.

Neil is a CAD expert, having many years of experience using numerous CAD systems, including AutoCAD, CADKEY/KeyCreator, Solid Edge, and of course SolidWorks.

For the record, Neil is Joe's nephew.

Tell Us What You Think!

As the reader of this book, you are our most important critic. We value your opinion and want to know how we are doing, good or bad. If you feel we missed something or could have done a better job, let us know. Also, if there are other areas of SolidWorks you feel need more explanation, tell us. We may be able to help.

You can email us at **books@SheetMetalGuy.com** to let us know what you did or didn't like about this book – as well as what we can do to make our books better.

When you write, please be sure to include the book's title as well as your name and contact information. We will carefully review your comments and share them with those whom helped make this book possible.

Table of Contents

Introduction

There is more to creating sheet metal parts than just adding flanges. Holes, patterns, and cutouts occur frequently and their creation is crucial to modeling these parts. The purpose of this book is to show you how to create the commonly used punching shapes and patterns.

The SolidWorks Design Library is a great feature which allows you to create and save special shapes to be recalled on later parts. You will learn how to make your own Design Library and fill it with your tools.

SolidWorks offers multiple methods of creating the same object. For example, you can create a Linear Pattern of a hole in a sketch or after the completion of a sketch. A round hole can be created using the Simple Hole command, the Cut-Extrude command, or the Hole Wizard. Each has its good and bad points when trying to achieve a specific goal with the hole definition. Throughout this book, you will use different methods of creation which we hope will provide you with the insight to make your own decisions about which method is the best for you.

You should be familiar with the part creation commands covered in Course 1. The part creation commands are fully included so that you can properly draw the parts. However, the instructions have been abbreviated since you should already know how to use these commands.

Conventions Used in this Book

It is assumed that you have a working knowledge of SolidWorks and the menu structure. You may want to open SolidWorks and in the "Help" menu, go through the **Online Tutorial**. In the first few chapters, we show the full CommandManager to help you learn what icons to select. Later chapters show only the icon to be selected. Dialog boxes, toolbars, and icons are shown in the book. When several icons appear in a dialog box, the one which you should select is circled in the picture in the book. A circle will not appear on your SolidWorks screen.

Setting the Toolbars to Match the Book

The CommandManager is a context-sensitive toolbar that dynamically updates based on the toolbar you want to access. By default, it has toolbars embedded in it based on the document type.

When you click a button in the control area, the CommandManager updates to show that toolbar. For example, if you click **Sketch** in the control area, the "Sketch" toolbar appears in the CommandManager.

Use the CommandManager to access toolbar icons in a central location and to save space for the graphics area.

To access the CommandManager, first open a new document. To do this, click the **New** icon in the "Standard" toolbar, or pull down the "File" menu and pick **New**.

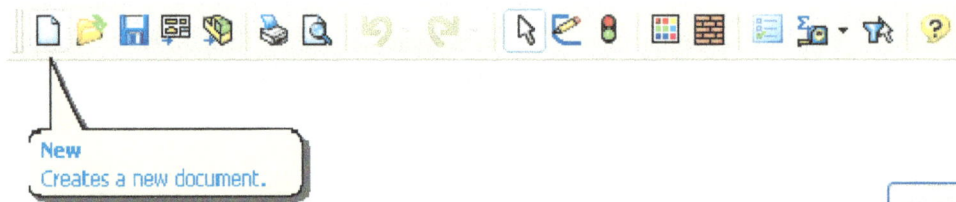

New
Creates a new document.

The **New SolidWorks Document** dialog box appears.

Click **Part** and then click **OK**. A new part window appears.

Pull down the "Tools" menu and pick **Customize**.

In the **Customize** dialog box on the **Toolbars** tab, make sure the **Enable CommandManager** check box is checked. If it is not, click the check box to check it. Then, select the **OK** button.

To make it easier for you to follow along and find the commands described in this book, you will want to make certain that the CommandManager is the same as displayed in the book.

To set up the CommandManager, move the cursor over the CommandManager and click the right mouse button. In the menu, make sure that **Show Description** is not checked. Then, pick **Customize Command Manager** as shown.

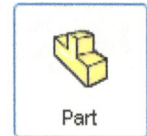

A long menu will appear, as shown to the right. Check **Features**, **Sheet Metal**, and **Sketch**. Make sure that all the others are unchecked. To accept your selections, simply click the left mouse button anywhere in the graphics area.

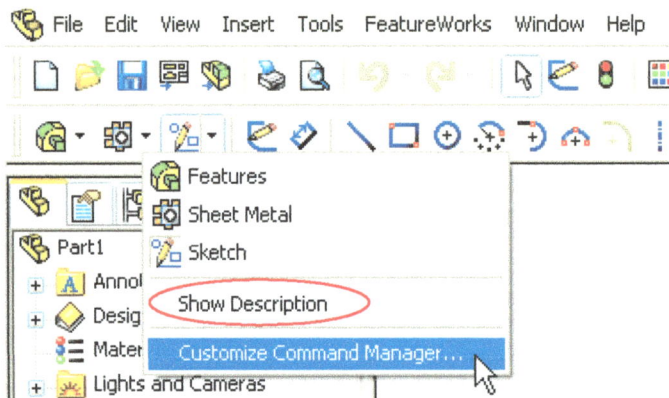

Part

- [ ] 2D to 3D
- [ ] Align
- [ ] Annotations
- [ ] Assemblies
- [ ] Blocks
- [ ] Curves
- [ ] Dimension/Relations
- [ ] Drawings
- [ ] Explode Sketch
- [ ] Fastening Feature
- [x] Features
- [ ] Fonts
- [ ] Line Formats
- [ ] Macros
- [ ] Molds
- [ ] Quick Snaps
- [ ] Reference Geometry
- [ ] Selection Filters
- [x] Sheet Metal
- [ ] Simulation
- [x] Sketch
- [ ] SolidWorks Office
- [ ] Splines
- [ ] Standard
- [ ] Standard Views
- [ ] Surfaces
- [ ] Tables
- [ ] Tools
- [ ] View
- [ ] Web
- [ ] Weldments

Chapter 1

End Cap

The End Cap starts with a series of line patterns of round holes. Since these holes are not intended to be centered on the length of the part, the Fill Pattern works well. When the part length changes, SolidWorks recalculates the Fill Pattern and updates the part automatically.

The sides of the part have a five hole pattern equally spaced. A unique solution shown here makes certain the spacing is always equal.

The holes on the end flanges are tied in dimensionally and mirrored to the opposite end so that future edits can be done once and both ends of the part are adjusted.

Create the Base Flange

Begin a new **Part** document by clicking the **New** icon in the "Standard" toolbar, or pull down the "File" menu and pick **New**.

Create a base flange by clicking the **Sheet Metal** icon in the control area of the CommandManager. Then, click the **Base-Flange/Tab** icon from the toolbar, or pull down the "Insert" menu and pick **Sheet Metal – Base Flange**.

Select the **Top** plane when prompted to select a plane on which to sketch the feature cross-section.

Create a rectangle with the origin inside the rectangle using the **Rectangle** icon in the CommandManager, or pull down the "Tools" menu and pick **Sketch Entities – Rectangle**.

Create a construction line diagonally across the rectangle by clicking the **Centerline** icon in the CommandManager, or pull down the "Tools" menu and pick **Sketch Entities – Centerline**.

Select the top left corner and then the bottom right corner of the rectangle that you just created to create the centerline as shown below. You will use this line to center the part on the origin. This will allow you to use the standard planes for mirroring rather than creating new planes.

Right click in the graphics area and pick **Select** from the menu. Select is the default mode when you are not in a command. You can use Select to exit the Centerline command and return to the Select mode, or you can press the **Escape** key on the keyboard

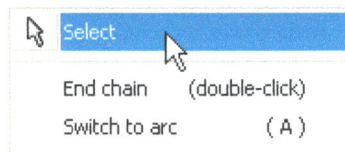

Select the diagonal line. Then, hold down the **Ctrl** key and select the origin.

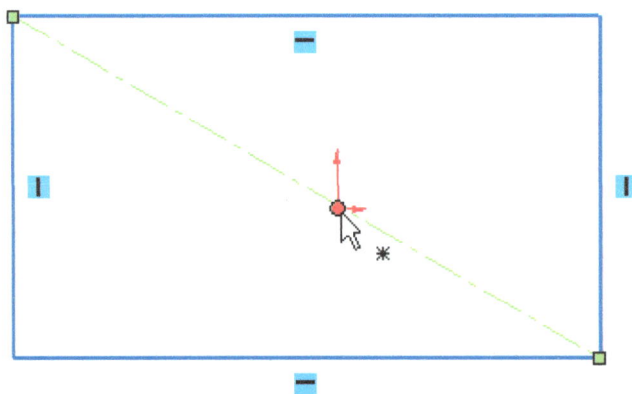

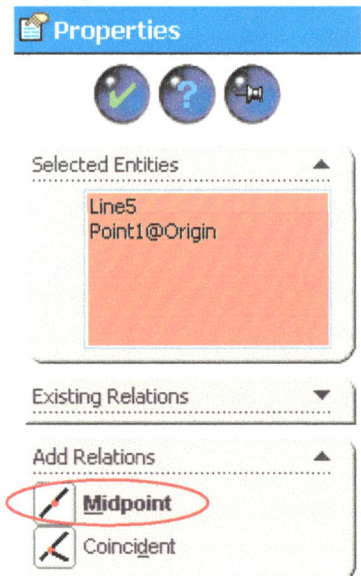

In the **Properties** PropertyManager, under **Add Relations**, click the **Midpoint** button.

✎ Add a '**2**' horizontal dimension to the bottom line and a '**10**' vertical dimension to the left vertical line using the **Smart Dimension** icon in the CommandManager, or pull down the "Tools" menu and pick **Dimensions – Smart**.

✏ Exit the sketch by clicking the **Exit Sketch** icon in the CommandManager or in the upper right corner of the graphics area.

↘T1 In the **Base Flange** PropertyManager under **Sheet Metal Parameters**, set the **Thickness** to '**.06**'. Make sure that the **Reverse direction** check box is not checked.

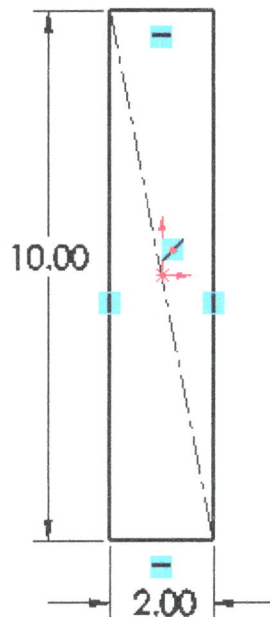

10.00

2.00

Sheet Metal Parameters
↘T1   0.06in
      ☐ Reverse direction

✅ Click the green check mark button at the top of the **Base Flange** PropertyManager to accept the settings and create the part.

Create the Edge Flanges

✎ Click the **Edge Flange** icon in the CommandManager, or pull down the "Insert" menu and pick **Sheet Metal – Edge Flange**.

Select the front edge of the base flange, move the cursor down, and click to set the direction of the flange.

Then, select the other three edges as shown.

In the **Edge-Flange** PropertyManager, set the **Flange Length** to **Blind** and the **Length** to '**1**'.

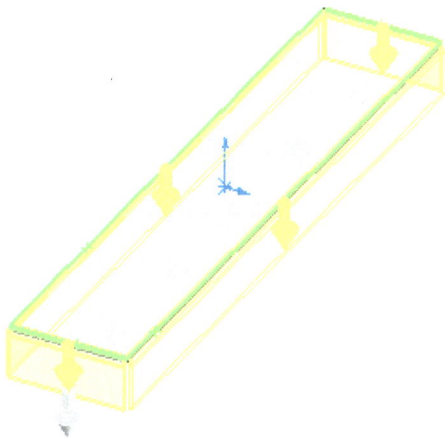

Click the **Outer Virtual Sharp** button and the **Material Inside Flange Position** button.

✅ Click the green check mark button at the top of the **Edge-Flange** PropertyManager to accept the settings and create the flange.

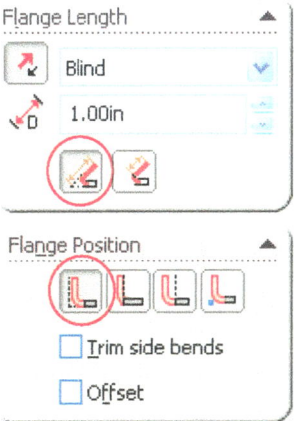

Flange Length
↗   Blind
↘D 1.00in

Flange Position
☐ Trim side bends
☐ Offset

Starting the Hole Pattern

Begin by clicking the **Extruded Cut** icon from the CommandManager or pull down the "Insert" menu and pick **Cut – Extrude**.

Extruded Cut
Cuts a solid model by extruding a sketched profile in one or two directions.

Select the top face of the End Cap as shown.

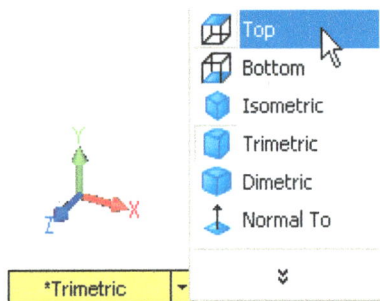

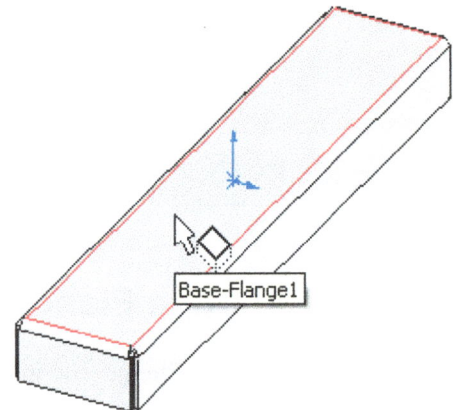

Base-Flange1

Top
Bottom
Isometric
Trimetric
Dimetric
Normal To

*Trimetric

In the bottom left corner of the graphics area, change the View orientation by clicking the pull down arrow and picking **Top**, or you can press and hold the **Ctrl** key and press the **5** key on the keyboard (**Ctrl-5**).

Create a vertical centerline line by clicking the **Centerline** icon in the CommandManager, or pull down the "Tools" menu and pick **Sketch Entities – Centerline**.

Select the midpoint of the top horizontal line followed by the midpoint of the bottom horizontal line.

Now, zoom in closer to the top of the part by clicking the **Zoom To Area** icon from the "View" toolbar. Make a box around the area that you want to zoom in to. A faster way to do this is to place the cursor over the area on the graphics screen that you want to zoom in or out on and scroll the middle mouse wheel, if you have one.

Zoom to Area
Zooms to the area you select with a bounding box.

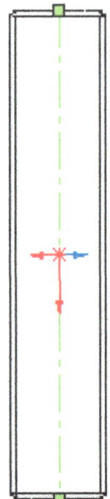

Create the three circles as shown below using the **Circle** icon in the CommandManager, or pull down the "Tools" menu and pick **Sketch Entities – Circle**. Make sure that the center point of the largest circle is on the centerline.

Right click in the graphics area and pick **Select** from the menu.

Select the center point of the large circle. Then, hold down the **Ctrl** key and select the center point of the small circle.

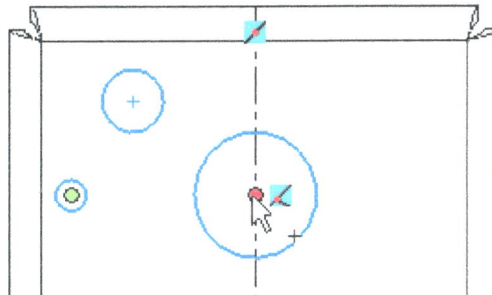

In the **Properties** PropertyManager, click the **Horizontal** button, and then click the green check mark button at the top of the **Properties** PropertyManager.

Right click in the graphics area and pick **Smart Dimension** from the menu. Add the dimensions as shown.

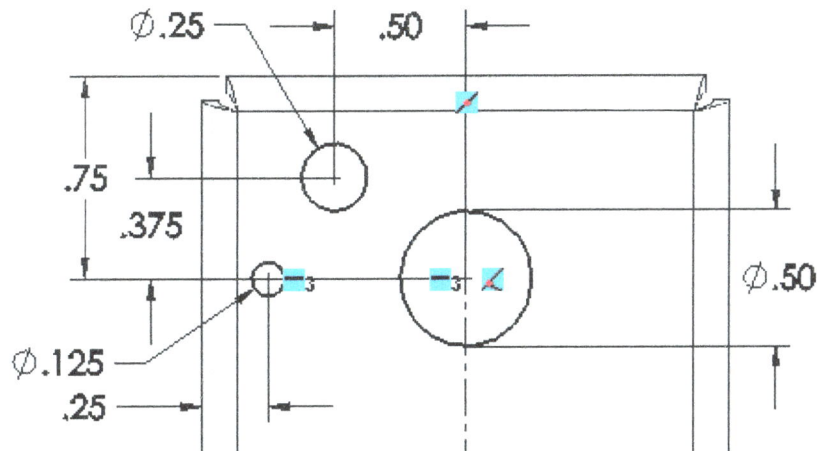

Click the **Mirror Entities** icon from the CommandManager, or pull down the "Tools" menu and pick **Sketch Tools – Mirror**.

Mirror Entities
Mirrors selected entities about a centerline.

 **Mirror**

Select the two smaller circles for the **Entities to Mirror**. Then, the **Mirror** PropertyManager, click in the **Mirror about** box under **Options**. Select the centerline as the entity to **Mirror about**.

Message

Select entities to mirror and a sketch line or linear model edge to mirror about

Options

Entities to mirror:

Arc2
Arc4

☑ Copy

Mirror about:

Line1

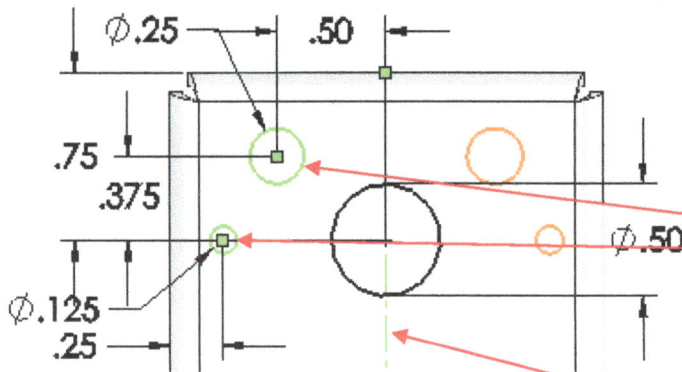

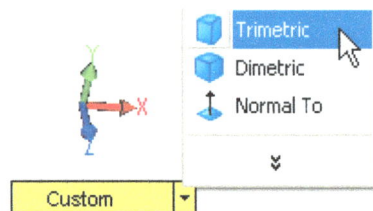 Click the green check mark button at the top of the **Mirror** PropertyManager.

Exit the sketch by clicking the **Exit Sketch** icon in the CommandManager or in the upper right corner of the graphics area.

In the **Cut-Extrude** PropertyManager, check the **Link to thickness** check box and then click the green check mark button at the top of the **Cut-Extrude** PropertyManager.

In the bottom left corner of your graphics area, click the pull down arrow and pick **Trimetric**.

Direction 1

Blind

☑ Link to thickness
☐ Flip side to cut
☑ Normal cut

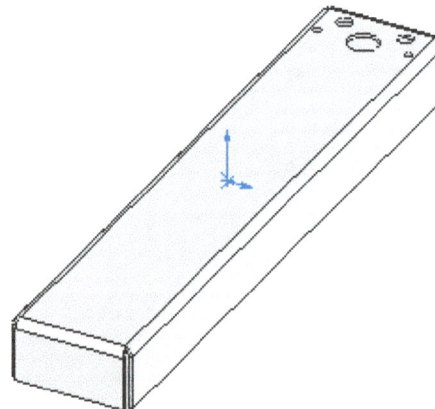

Rename the Feature

In the FeatureManager design tree, slowly click two times (left click once, pause, left click again) on **Cut-Extrude1** to select it, or you can click once on the feature name and then press the **F2** key on the keyboard.

Then, enter '**Top Holes**' for the new name and press **Enter** to accept it.

Sheet-Metal1
Base-Flange1
Edge-Flange1
Top Holes
Flat-Pattern1

Finishing the Hole Pattern

Pull down the "Insert" menu and pick **Pattern/Mirror - Fill Pattern**.

In the graphics area, select the top face of the End Cap for the **Fill Boundary**.

In the **Fill Pattern** PropertyManager under **Pattern Layout**, set the **Instance Spacing** to '**.625**'.

If under **Features to Pattern** you do not see **Top Holes**, click in the **Features to Pattern** box. Then, in the upper left hand corner of the graphics area, click on the plus sign next to the document name (Part1) to expand the flyout FeatureManager design tree. You may also click on the PropertyManager Title Block (where it says **Fill Pattern**) to expand the flyout FeatureManager design tree.

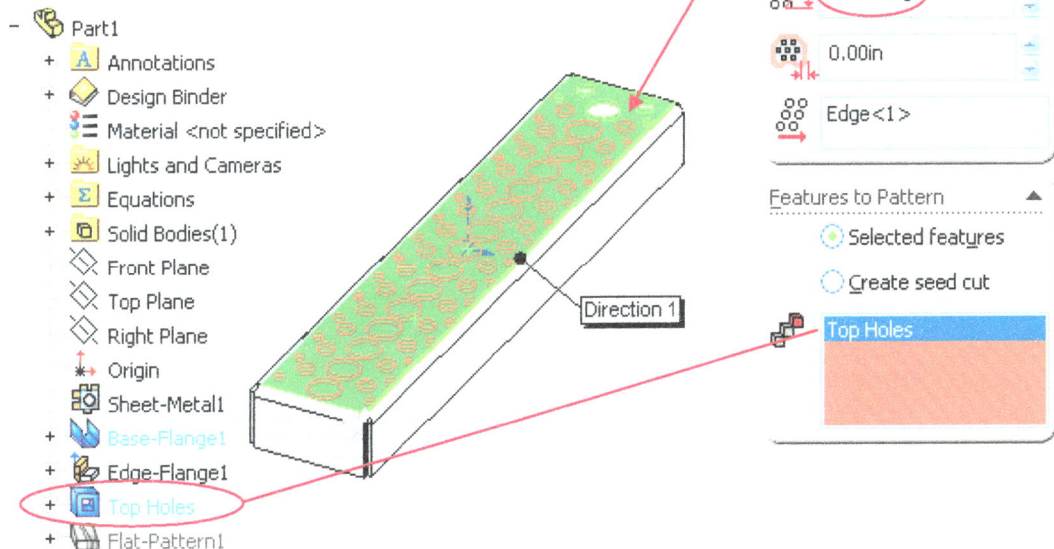

Under **Features to Pattern**, select **Top Holes** from the flyout FeatureManager design tree.

Click the green check mark button at the top of the **Fill Pattern** PropertyManager.

Add Holes on the Side Flange

To do this, click the **Extruded Cut** icon from the CommandManager or pull down the "Insert" menu and pick **Cut – Extrude**.

Select the right side of the part for the plane to sketch on.

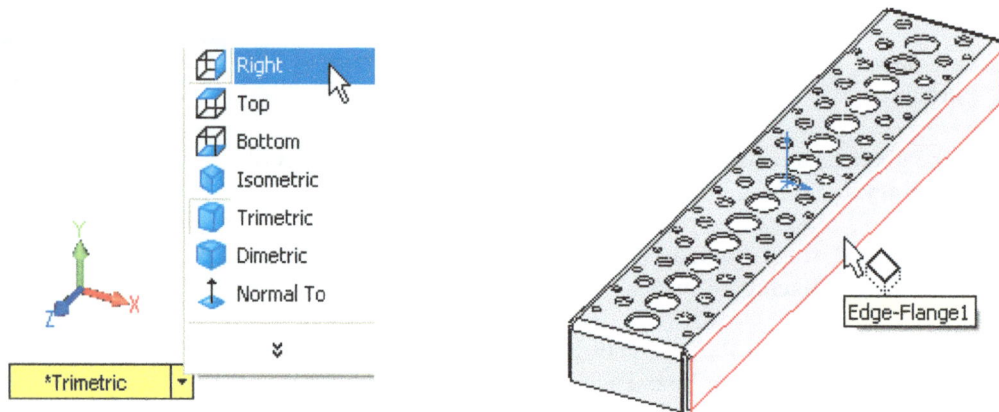

In the bottom left corner of your graphics area, click the pull down arrow and pick **Right**, or you can press **Ctrl-4** on the keyboard.

Create a vertical centerline line by clicking the **Centerline** icon in the CommandManager, or pull down the "Tools" menu and pick **Sketch Entities – Centerline**.

Place the centerline through the origin of the part.

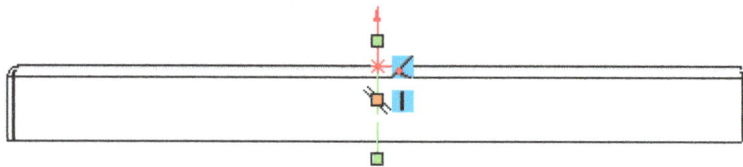

Create a circle with its center point on the centerline using the **Circle** icon in the CommandManager, or pull down the "Tools" menu and pick **Sketch Entities – Circle**.

Create two more circles, one near the left edge and another near the right edge.

Right click in the graphics area and pick **Select** from the menu or press the **Escape** key.

Hold down the **Ctrl** key and select the center points of the three circles. Make sure that the cursor shows that you are on the center point and that the circle is not selected.

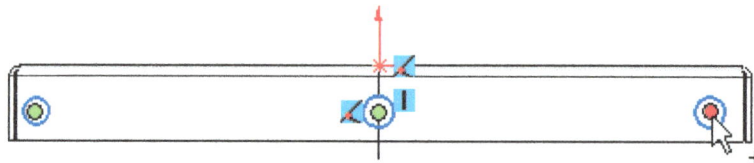

In the **Properties** PropertyManager, click the **Horizontal** button, and then click the green check mark button at the top of the **Properties** PropertyManager.

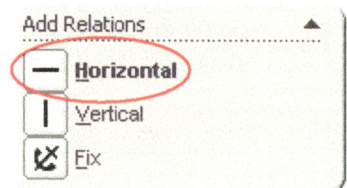

SolidWorks for the Sheet Metal Guy

Create a centerline from the center of the leftmost circle to the center of the middle circle.

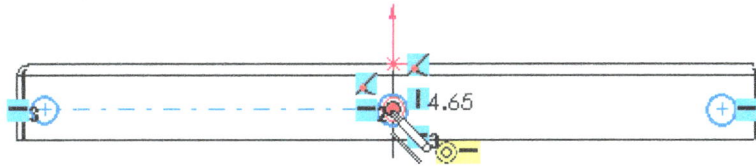

Create another centerline from the center of the middle circle to the center of the rightmost circle.

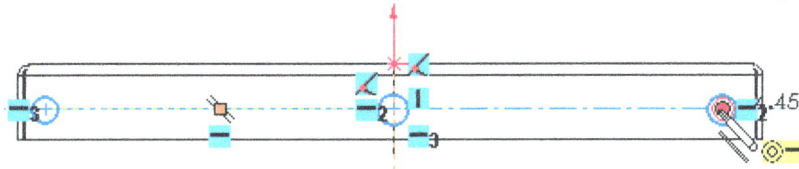

Create a circle at the midpoint of each horizontal centerline.

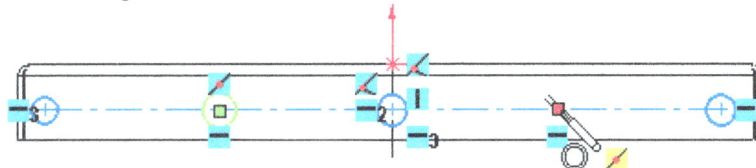

Right click in the graphics area and pick **Select** from the menu or press the **Escape** key.

Select the leftmost circle. Hold down the **Ctrl** key and select the other four circles. This time, make sure that you select the circles and not the center points.

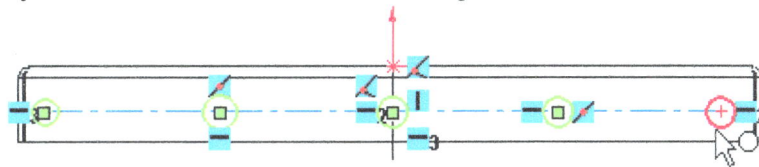

In the **Properties** PropertyManager, click the **Equal** button.

Click the **Smart Dimension** icon and add a '.25' dimension to one of the circles.

Next, add a '.5' dimension from the leftmost circle to the outside edge of the left end flange. Add a similar '.5' dimension to the rightmost circle as shown.

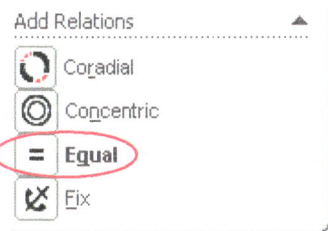

Finally, add a '.4375' dimension from the bottom edge to the horizontal centerline.

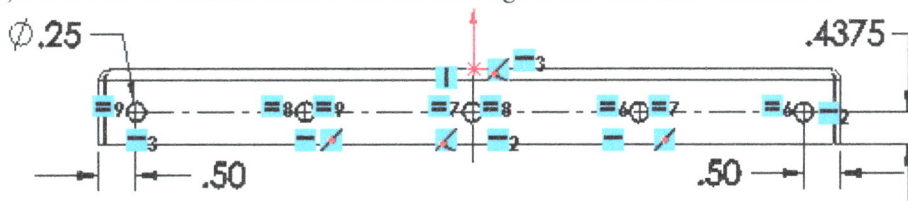

Exit the sketch by clicking the **Exit Sketch** icon in the CommandManager or in the upper right corner of the graphics area.

In the **Cut-Extrude** PropertyManager, check the **Link to thickness** check box and then click the green check mark button at the top of the **Cut-Extrude** PropertyManager.

Add Holes to the Front Flange

For the holes on the front of the part, holes will be created at a set distance from the side of the part.

In the bottom left corner of your graphics area, click the pull down arrow and pick **Front**.

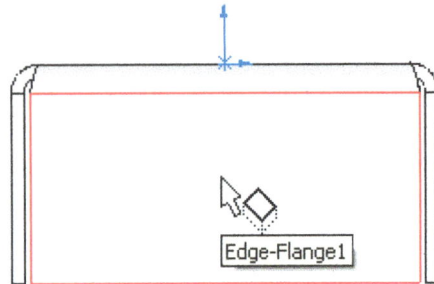

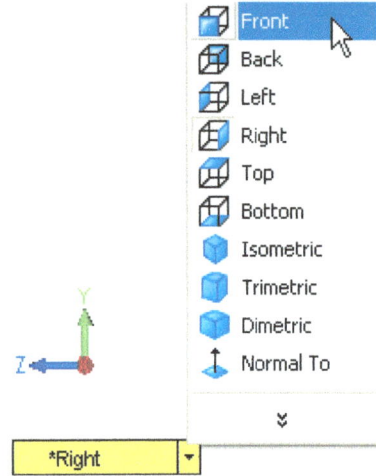

Edge-Flange1

Click the **Extruded Cut** icon from the CommandManager or pull down the "Insert" menu and pick **Cut – Extrude**.

*Right

Extruded Cut
Cuts a solid model by extruding a sketched profile in one or two directions.

Direction 1
Blind
☑ Link to thickness
☐ Flip side to cut
☑ Normal cut

Front
Back
Left
Right
Top
Bottom
Isometric
Trimetric
Dimetric
Normal To

Select the front of the part for the plane to sketch on as shown above.

Create a horizontal centerline line by clicking the **Centerline** icon in the CommandManager, or pull down the "Tools" menu and pick **Sketch Entities – Centerline**. Make sure that you do not select the midpoint of any of the vertical lines.

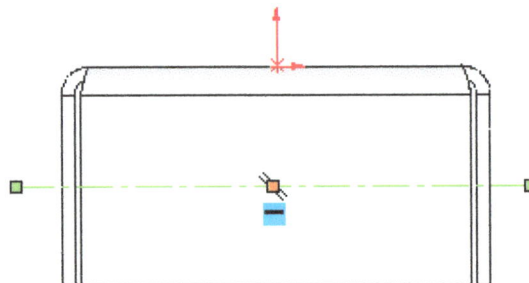

Create one circle with its center point on the centerline near the left edge and another circle with its center point on the centerline near the right edge using the **Circle** icon in the CommandManager, or pull down the "Tools" menu and pick **Sketch Entities – Circle**.

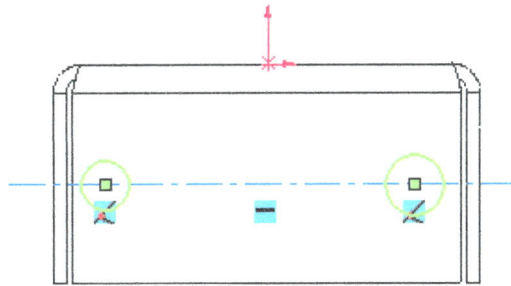

Right click in the graphics area and pick **Select** from the menu or press the **Escape** key.

Select the leftmost circle. Hold down the **Ctrl** key and select the other circle.

In the **Properties** PropertyManager, click the **Equal** button.

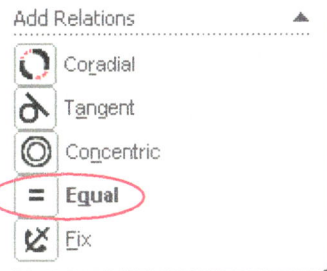

Click the **Smart Dimension** icon and add a '**.25**' dimension to the left circle. Then, add '**.375**' dimensions from the edge of the part to the center of the circles. Finally, add a '**.4375**' dimension from the bottom edge to the horizontal centerline.

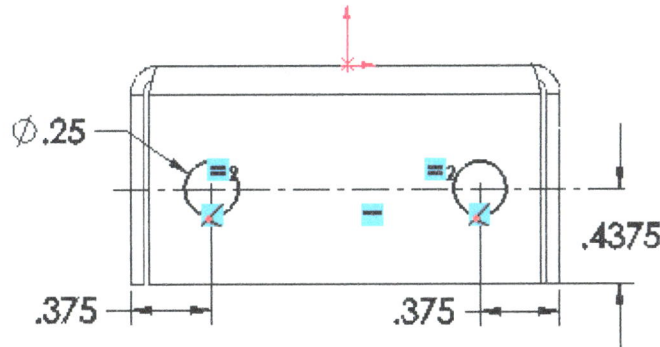

Exit the sketch by clicking the **Exit Sketch** icon in the CommandManager or in the upper right corner of the graphics area.

In the **Cut-Extrude** PropertyManager, check the **Link to thickness** check box and then click the green check mark button at the top of the **Cut-Extrude** PropertyManager.

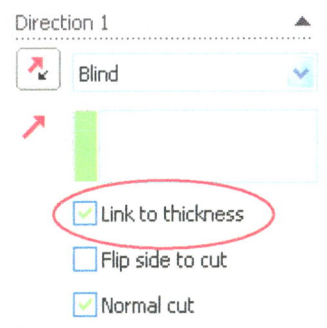

In the bottom left corner of your graphics area, click the pull down arrow and pick **Trimetric**.

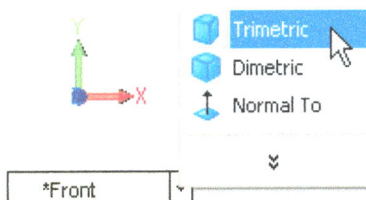

Rename the Features

In the FeatureManager design tree, slowly click two times (left click once, pause, left click again) on **Cut-Extrude2** to select it.

Then, enter '**Side Holes**' for the new name. Do the same for **Cut-Extrude3** and enter '**Front Holes**' for the new name. Note that when you select an feature from the FeatureManager design tree SolidWorks highlights the geometry in the graphics area, letting you know that you selected the correct feature.

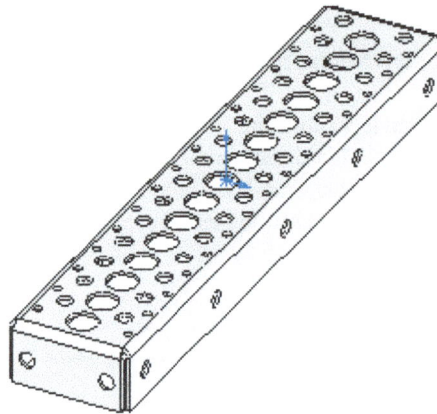

Press the **Escape** key to return to select mode and deselect any selected features.

Mirror the Holes

Click the **Features** icon in the control area of the CommandManager. Then, click the **Mirror** icon from the toolbar, or pull down the "Insert" menu and pick **Pattern/Mirror – Mirror**.

Mirror
Mirrors features, faces, and bodies about a face or a plane.

In the upper left hand corner of the graphics area, click on the plus sign next to the document name (Part1) to expand the flyout FeatureManager design tree.

In the flyout FeatureManager design tree, select **Right Plane** as the **Mirror Face/Plane**.

Under **Features to Mirror**, select **Side Holes** from the flyout FeatureManager design tree.

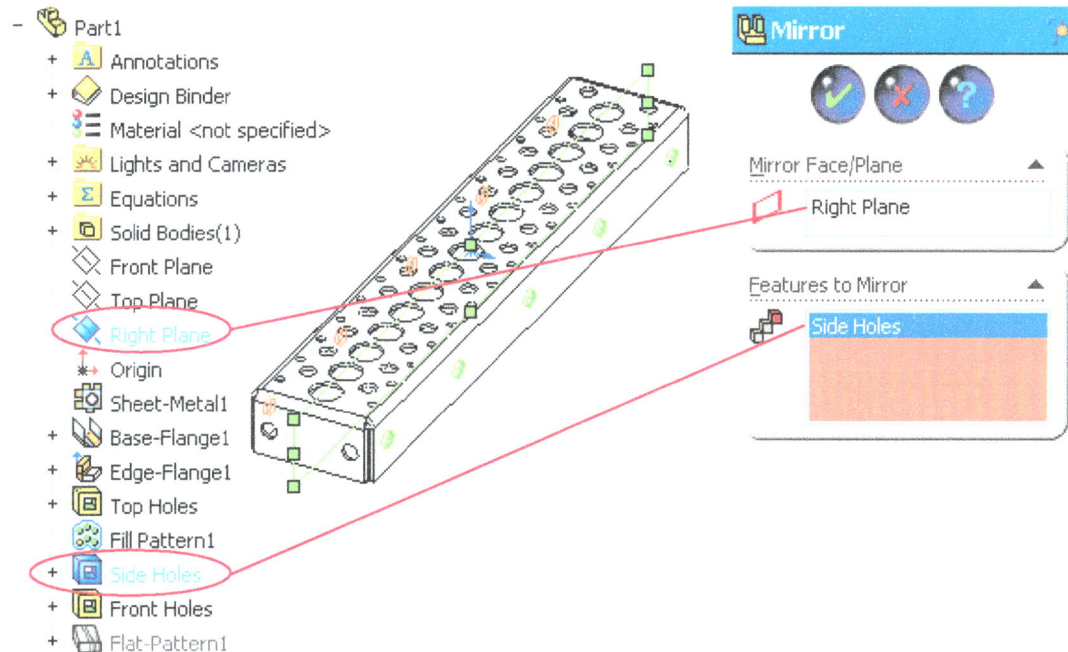

Click the green check mark button at the top of the **Mirror** PropertyManager.

Press the **Escape** key to deselect any selected features.

Click the **Mirror** icon again, but this time, select **Front Plane** as the **Mirror Face/Plane,** and under **Features to Mirror**, select **Front Holes**.

If **Mirror1** is already under **Features to Mirror**, simply select it in the **Mirror** PropertyManager and press the **Delete** key.

Click the green check mark button at the top of the **Mirror** PropertyManager.

Rotate the part around so that you can see all the holes.

In the bottom left corner of your graphics area, click the pull down arrow and pick **Trimetric**.

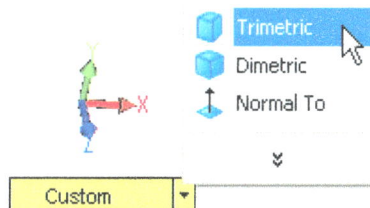

Save the Part

Click the **Save** icon on the "Standard" toolbar, or pull down the "File" menu and pick **Save**.

The **Save As** dialog box appears. Make certain that you are in the desired folder. In the **File name** box, type '**End Cap**' and click **Save**.

Changing the Part

Now, no matter what the size of the part, the pattern of the five holes will fill the entire top of the part. The holes on the side and end of the part, will maintain their correct locations.

For example, double click on **Base-Flange1** in the FeatureManager design tree. Then, double click on the **10.00** dimension and change the value to '**17.5**'.

Click the **Rebuild** icon in the "Standard" toolbar to see the changes, or press the keyboard shortcut **Ctrl+B** to update the model.

The pattern on the top of the part has updated. Also, notice that the holes on the side of the part still meet the design requirements.

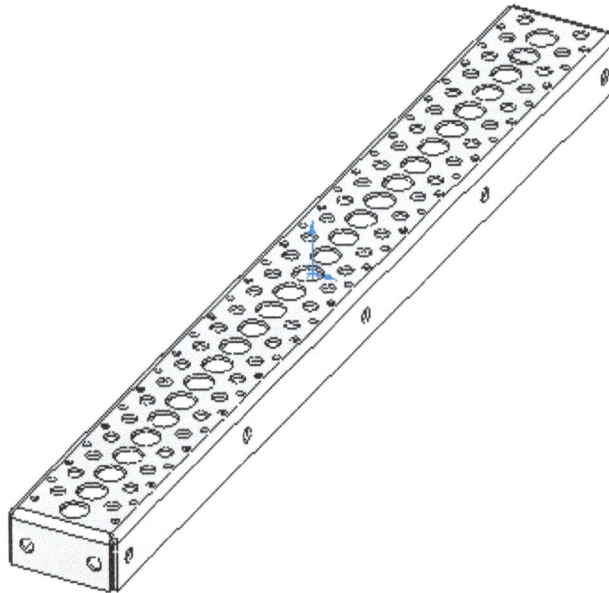

Closing the File

Pull down the "File" menu and pick **Close**.

Click **No** when prompted to **Save changes to End Cap.SLDPRT?** to not save your changes.

Chapter 2

Housing Bracket

The Housing Bracket provides another example of the Fill Pattern. This time the holes are centered on the flange.

The tabs are created by modeling one of them, including the obround hole, and mirror imaging the tab and the hole to the other side of the part. When mirroring features on a symmetrical part, it is very helpful to center the origin on the part as was done in this lesson.

The special shape cutouts are then added to the base of the part. Again a Mirror command is used to allow you to create the shape once and maintain the two identical features.

Create the Base Flange

Begin a new **Part** document by clicking the **New** icon in the "Standard" toolbar, or pull down the "File" menu and pick **New**.

Create a base flange by clicking the **Sheet Metal** icon in the control area of the CommandManager. Then, click the **Base-Flange/Tab** icon from the toolbar, or pull down the "Insert" menu and pick **Sheet Metal – Base Flange**.

Select the **Top** plane when prompted to select a plane on which to sketch the feature cross-section.

Create a rectangle with the origin inside the rectangle using the **Rectangle** icon in the CommandManager, or pull down the "Tools" menu and pick **Sketch Entities – Rectangle**.

Create a construction line diagonally across the rectangle by clicking the **Centerline** icon in the CommandManager, or pull down the "Tools" menu and pick **Sketch Entities – Centerline**.

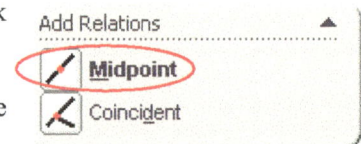

Select the top left corner and then the bottom right corner to create the centerline.

Press the **Escape** key on the keyboard to deselect the **Centerline** tool.

Select the diagonal line. Then, hold down the **Ctrl** key and select the origin.

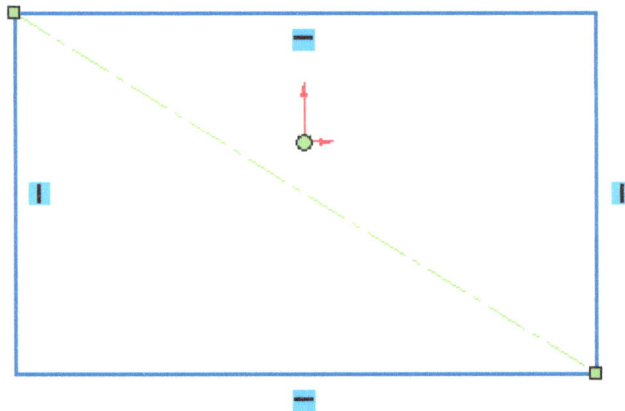

In the **Properties** PropertyManager, under **Add Relations**, click the **Midpoint** button.

Click the **Line** icon in the CommandManager, or pull down the "Tools" menu and pick **Sketch Entities – Line**.

Create the six lines as shown below to create the notches.

Next, click the **Trim Entities** icon or pull down the "Tools" menu and pick **Sketch Tools - Trim Entities**.

Trim to closest

Trim Entities
Trims or extends a sketch entity to be coincident to another, or deletes a sketch entity.

In the **Trim** PropertyManager, make sure that the **Trim to closest** button is depressed, and select the middle section of the top and bottom horizontal lines to trim out the notch.

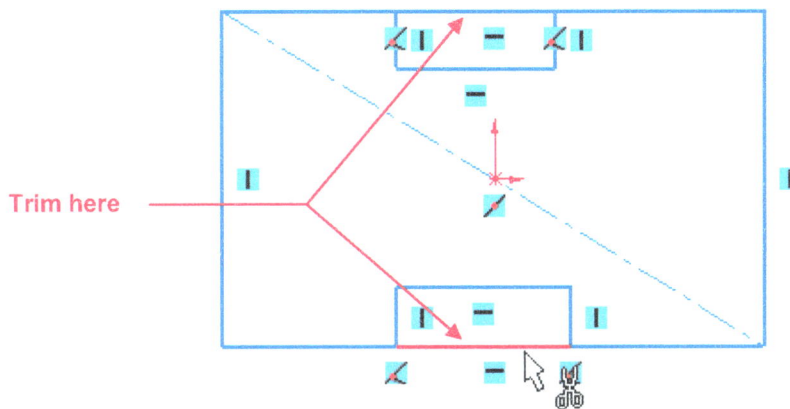

Trim here

Press the **Escape** key on the keyboard to deselect the **Trim** tool.

Select top left horizontal line. Hold down the **Ctrl** key and select the other three horizontal lines as shown.

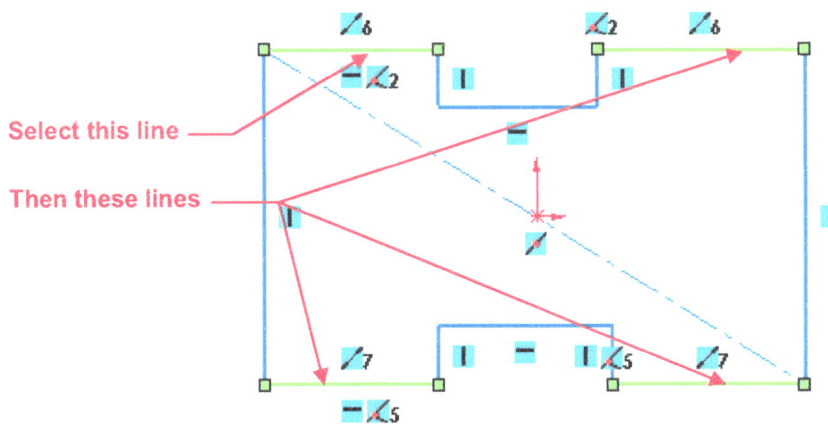

Select this line

Then these lines

In the **Properties** PropertyManager, click the **Equal** button.

Then, select the top left small vertical line. Hold down the **Ctrl** key and select the other three small vertical lines as shown.

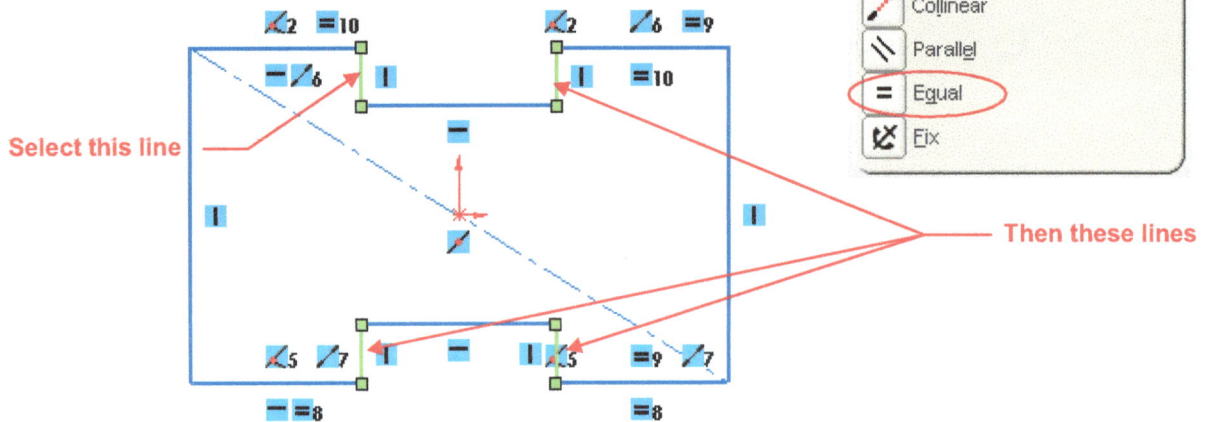

In the **Properties** PropertyManager, click the **Equal** button.

Click the **Smart Dimension** icon in the CommandManager, or pull down the "Tools" menu and pick **Dimensions – Smart**.

Add a '**6.00**' vertical dimension to the left vertical line and a '**12.00**' horizontal dimension between the left and right side of the sketch.

Press the **F** key on the keyboard to **Zoom to Fit** so you can see the entire rectangle and center it in the graphics area. (Keyboard shortcut key, **Zoom to Fit**: f)

Then, add a '**1.00**' vertical dimension for the depth of the notch and a '**2.00**' horizontal dimension for the width of the notch.

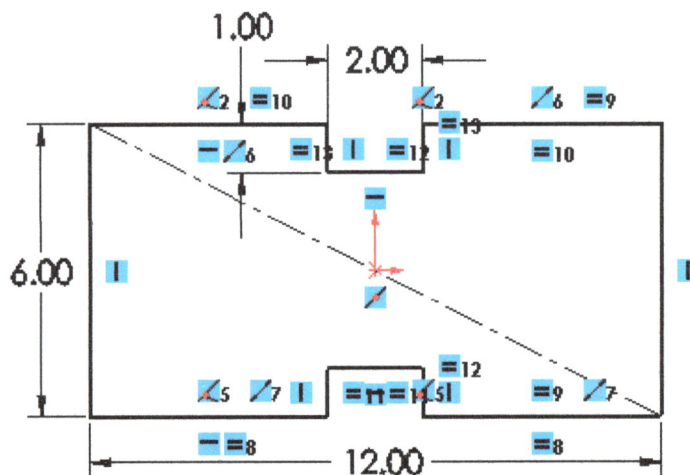

Exit the sketch by clicking the **Exit Sketch** icon in the CommandManager or in the upper right corner of the graphics area.

In the **Base Flange** PropertyManager under **Sheet Metal Parameters**, set the **Thickness** to '**.048**'. Make sure that the **Reverse direction** check box is checked.

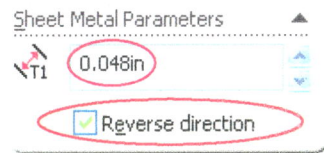

Click the green check mark button at the top of the **Base Flange** PropertyManager to accept the settings and create the part.

Create the Edge Flanges

Click the **Edge Flange** icon in the CommandManager, or pull down the "Insert" menu and pick **Sheet Metal – Edge Flange**.

Select the front left edge of the base flange and move the cursor up and click to set the direction of the flange.

Then, select the other three edges as shown.

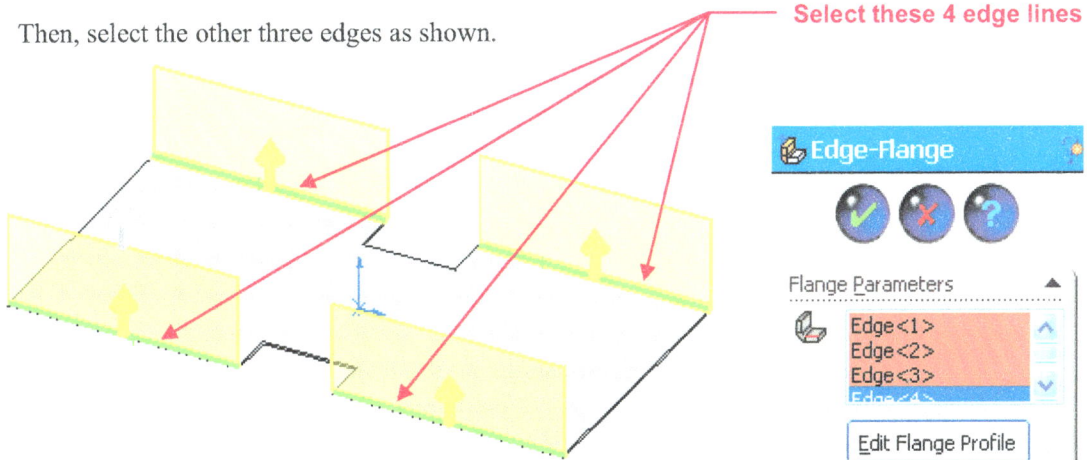

In the **Edge-Flange** PropertyManager, make sure that the Bend Radius is set to '**.075**'.

Set the **Flange Length** to **Blind** and the **Length** to '**2**'.

Click the **Outer Virtual Sharp** button and the **Material Inside Flange Position** button.

Click the green check mark button at the top of the **Edge-Flange** PropertyManager to accept the settings and create the flange.

Click the **Edge Flange** icon again in the CommandManager, or pull down the "Insert" menu and pick **Sheet Metal – Edge Flange**.

Select the front left edge of the edge flange and move the cursor to the right and click to set the direction of the flange.

Then, select the other three edges as shown.

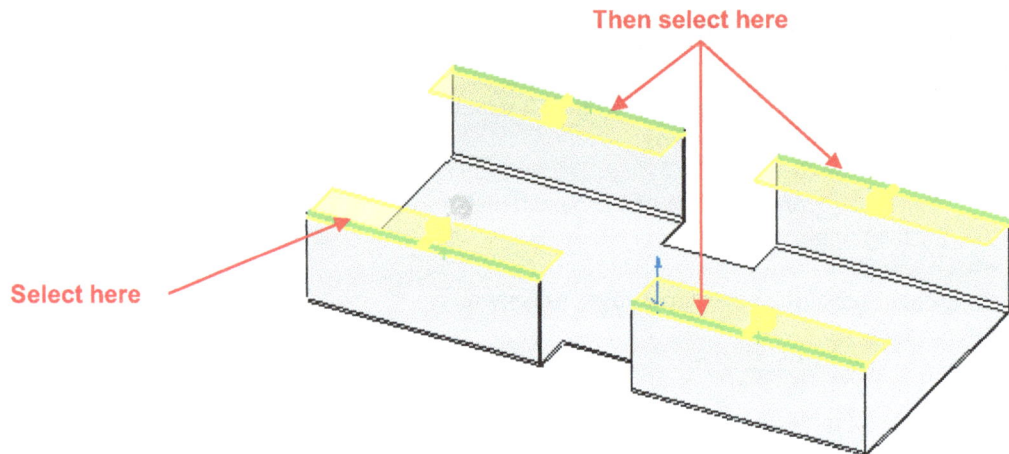

Then select here

Select here

In the **Edge-Flange** PropertyManager, set the **Flange Length** to **Blind** and the **Length** to '1'.

Click the **Outer Virtual Sharp** button and the **Material Inside Flange Position** button.

Click the green check mark button at the top of the **Edge-Flange** PropertyManager to accept the settings and create the flange.

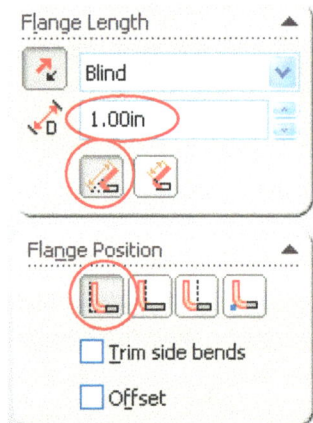

Flange Length

Blind

1.00in

Flange Position

Trim side bends

Offset

Create a Fill Pattern

Before creating the fill pattern, you must first create a feature to seed the pattern.

Click the **Simple Hole** icon in the CommandManager, or pull down the "Insert" menu and select **Features – Hole – Simple**.

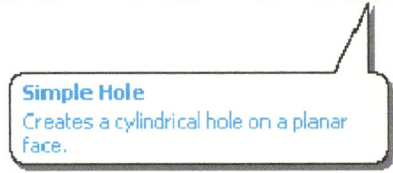

Simple Hole
Creates a cylindrical hole on a planar face.

When prompted in the PropertyManager to select a location for the center of the hole, pick the front of the edge flange as shown.

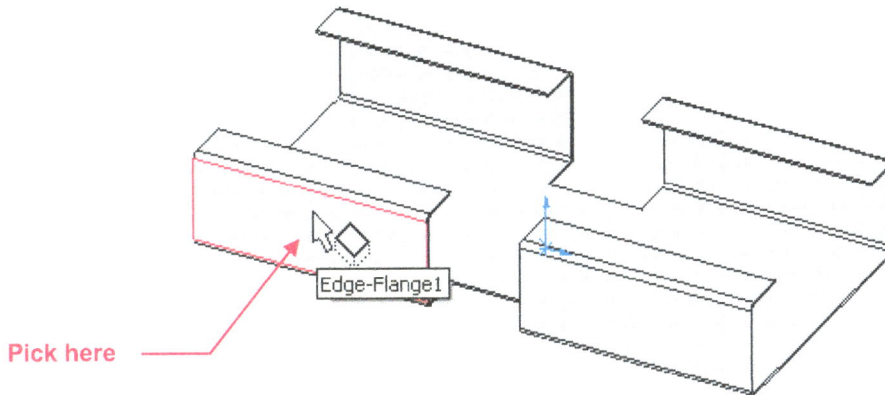

Pick here

Edge-Flange1

In the **Hole** PropertyManager, check the **Link to thickness** check box. This option ensures that the hole is through the thickness of the sheet metal, no matter what gauge or thickness is specified.

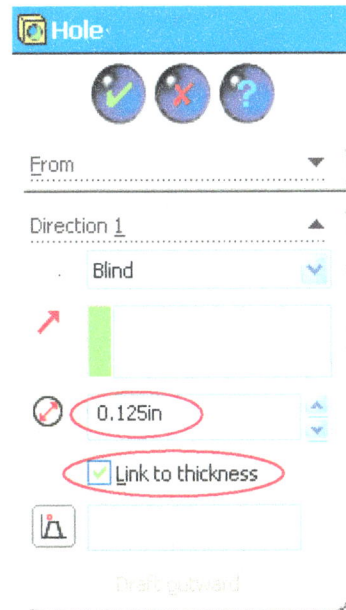

Enter '**.125**' for the diameter.

Click the green check mark button to accept the settings and create the hole.

Since the hole is placed at the cursor location, you need to edit the sketch in order to locate the hole.

In the FeatureManager design tree, right click on **Hole1** and pick **Edit Sketch**.

Change the display view to **Normal To** the sketch by picking **Normal To** in the View list in the lower left corner of the graphics area.

Trimetric
Dimetric
Normal To

*Trimetric

Click the **Centerline** icon in the CommandManager, or pull down the "Tools" menu and pick **Sketch Entities – Centerline**.

Select the top left corner and then the bottom right corner of the left side of the part to create the centerline.

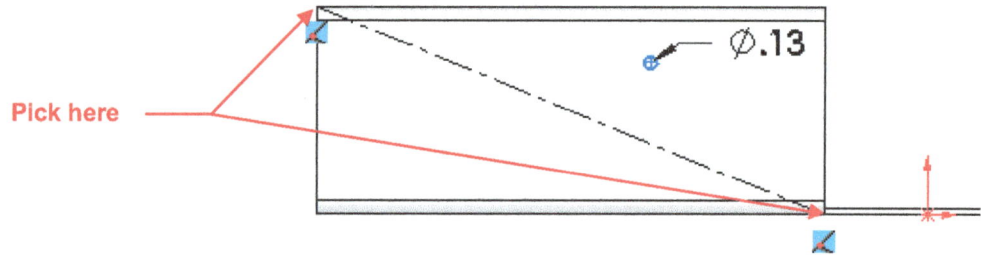

Press the **Escape** key on the keyboard to deselect the **Centerline** tool.

Left click and drag the center point of the circle to the midpoint of the centerline. Make sure that the Midpoint relation flag appears. You may need to drag the .13 dimension out of the way in order to select the circle centerpoints. You may also want to zoom in closer to ensure that the correct points are selected.

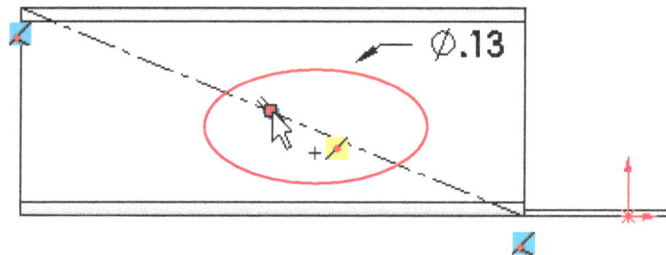

Exit the sketch by clicking the **Exit Sketch** icon in the CommandManager or in the upper right corner of the graphics area.

Pull down the "Insert" menu and pick **Pattern/Mirror - Fill Pattern**.

In the graphics area, select both flanges of the front of the part as shown. SolidWorks requires coplanar faces or contours, which these are.

In the **Fill Pattern** PropertyManager, set the **Spacing** to '**.25**', set the **Stagger Angle** to '**45**', and set the **Margins** to '**.1**'.

Click in the arrow next to **Features to Pattern** to expand the feature. Then, click in the **Features to Pattern** box.

In the upper left hand corner of the graphics area, click on the plus sign next to the document name (Part1) to expand the flyout FeatureManager design tree. In the flyout FeatureManager design tree, select **Hole1**. Since the hole depth is linked to the part thickness, the fill pattern will only cut through the selected flanges.

Click the green check mark button at the top of the **Fill Pattern** PropertyManager.

Click the View orientation pull down arrow in the bottom left corner of your graphics area and pick **Trimetric**.

Mirror the Hole

Press the **Escape** key to make sure that nothing is selected.

Click the **Features** icon in the control area of the CommandManager. Then, click the **Mirror** icon from the toolbar, or pull down the "Insert" menu and pick **Pattern/Mirror – Mirror**.

Mirror
Mirrors features, faces, and bodies about a face or a plane.

In the flyout FeatureManager design tree, select **Front Plane** for the **Mirror Face/Plane**.

In the **Mirror** PropertyManager, click in the **Features to Mirror** box. Select **Hole1** in the flyout FeatureManager design tree.

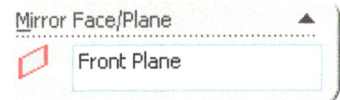

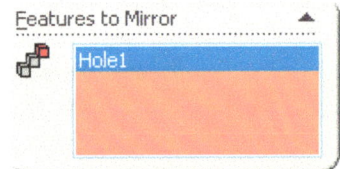

Click the green check mark button at the top of the **Mirror** PropertyManager.

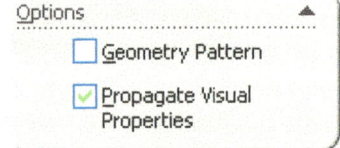

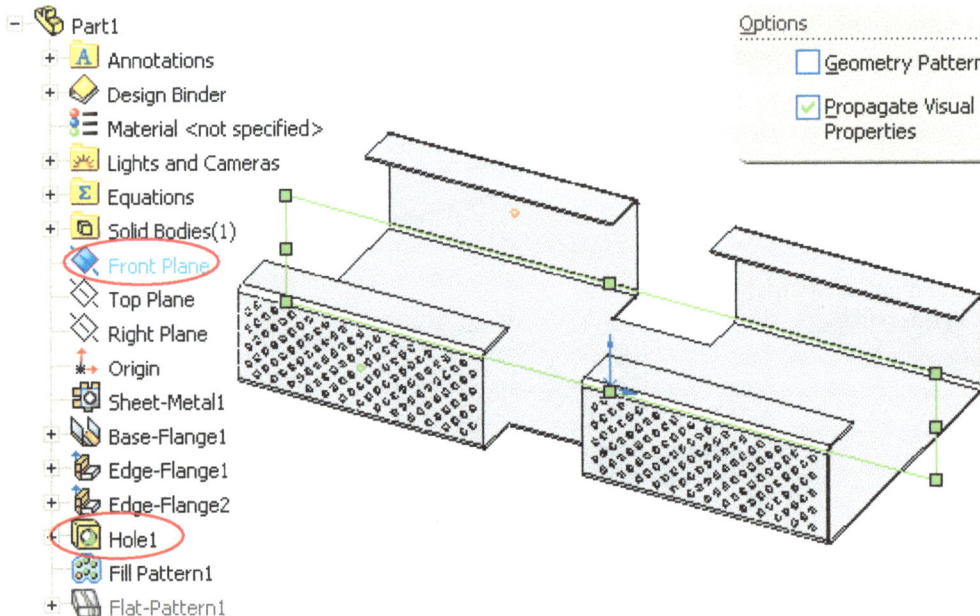

SolidWorks for the Sheet Metal Guy

Change the display view to **Back** by picking **Back** in the View list in the lower left corner of the graphics area.

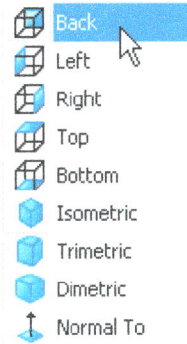

Add Another Fill Pattern

Pull down the "Insert" menu and pick **Pattern/Mirror - Fill Pattern**.

In the graphics area, select the both sides of the back of the part as shown.

In the **Fill Pattern** PropertyManager, set the **Spacing** to '**.25**', set the **Stagger Angle** to '**45**', and set the **Margins** to '**.1**'.

Click in **Features to Pattern**. Then, in the flyout FeatureManager design tree, select **Mirror1**.

If the size of Hole1 ever changes, filling with Mirror1 ensures that both sides will remain equal.

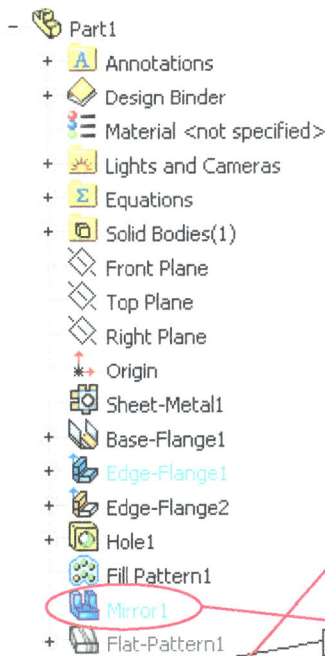

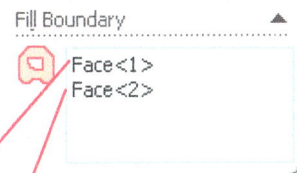

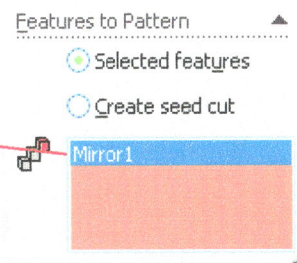

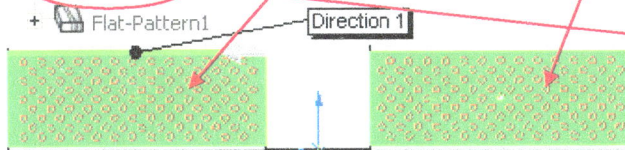

Change back to the **Trimetric** view by clicking the View orientation pull down arrow in the bottom left corner of your graphics area and picking **Trimetric** to see the Fill Pattern complete on both sides of the part.

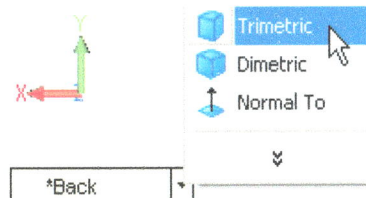

Create the Tabs

Click the **Sheet Metal** icon in the control area of the CommandManager. Then, click the **Edge Flange** icon in the CommandManager, or pull down the "Insert" menu and pick **Sheet Metal – Edge Flange**.

Select the front edge of the base flange as shown and move the cursor down. Click to set the direction of the flange.

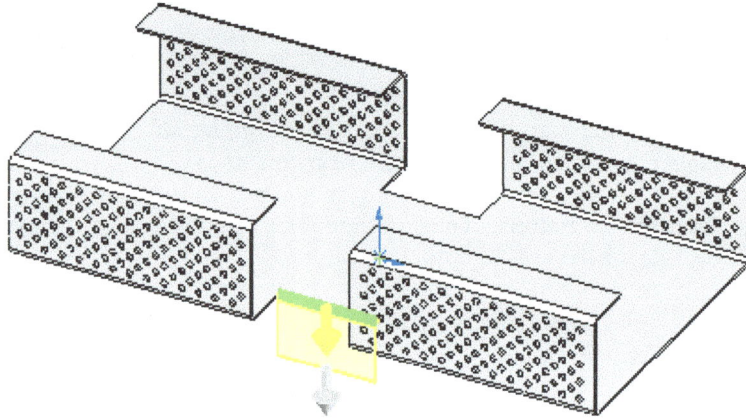

In the **Edge-Flange** PropertyManager, click on the **Edit Flange Profile** button.

Click the View orientation pull down arrow in the bottom left corner of your graphics area and pick **Front**.

Click and drag the left and right side of the sketch rectangle towards the middle of the sketch as shown.

Click the **Smart Dimension** icon in the CommandManager, or pull down the "Tools" menu and pick **Dimensions – Smart.**

Add a '**.25**' horizontal dimension between the left vertical sketch line and the point as shown below. You may need to zoom into the area to see it better.

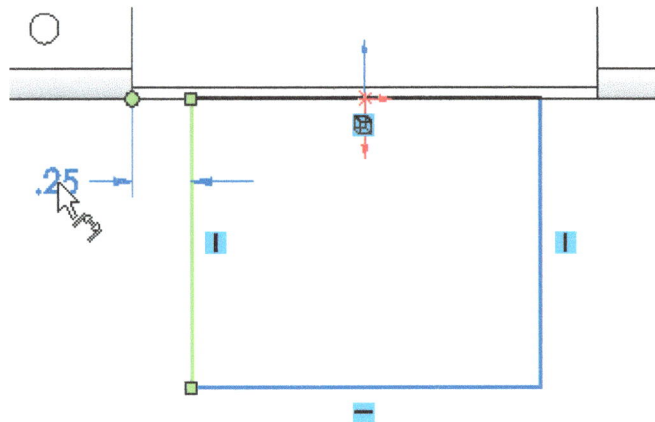

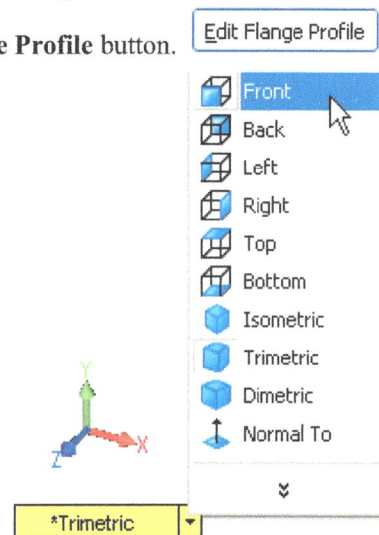

SolidWorks for the Sheet Metal Guy

Add another '.25' horizontal dimension on the other side between the right vertical sketch line and the opposite point.

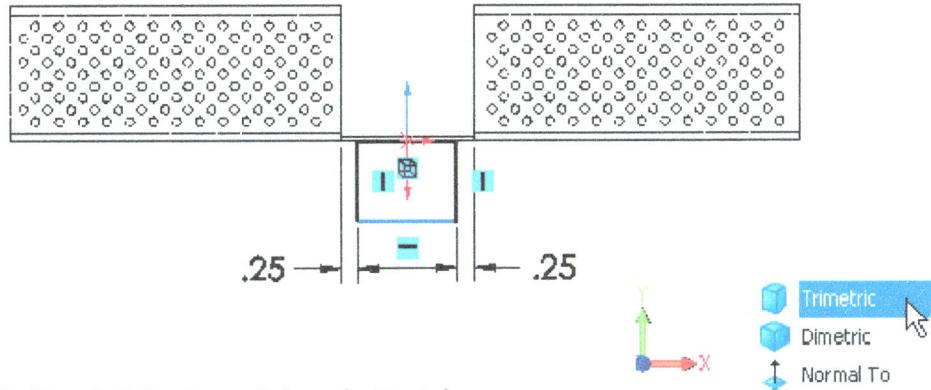

.25 —⊢⊣— .25

In the **Profile Sketch** dialog box, click on the **Back** button.

Click the View orientation pull down arrow in the bottom left corner of your graphics area and pick **Trimetric**.

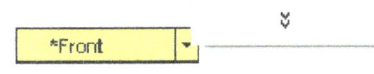

In the **Edge-Flange** PropertyManager, set the **Flange Length** to **Blind** and the **Length** to '**1.25**'.

Click the **Outer Virtual Sharp** button and the **Material Outside Flange Position** button.

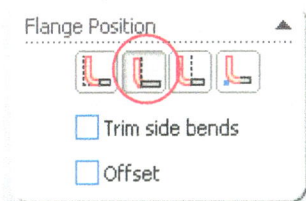

Click on the **Custom Relief Type** check box to use a custom relief.

Set the **Relief Type** to **Obround**.

Uncheck **Use relief ratio**. Set the **Relief Width** to '**.25**' and the **Relief Depth** to '**0**'.

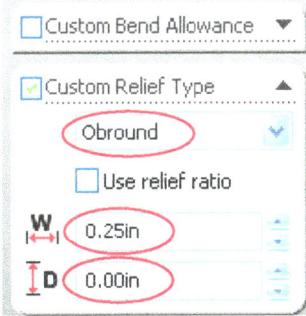

Click the green check mark button at the top of the **Edge-Flange** PropertyManager to accept the settings and create the flange.

Rename the Feature

In the FeatureManager design tree, slowly click two times (left click once, pause, left click again) on **Edge-Flange3** to select it.

Then, enter '**Tab**' for the new name.

Obround Hole

Click the **Sheet Metal** icon in the control area of the CommandManager. Then, click the **Extruded Cut** icon from the CommandManager or pull down the "Insert" menu and pick **Cut – Extrude**.

Extruded Cut
Cuts a solid model by extruding a
sketched profile in one or two directions.

Select the front of the tab for the plane to sketch on as shown.

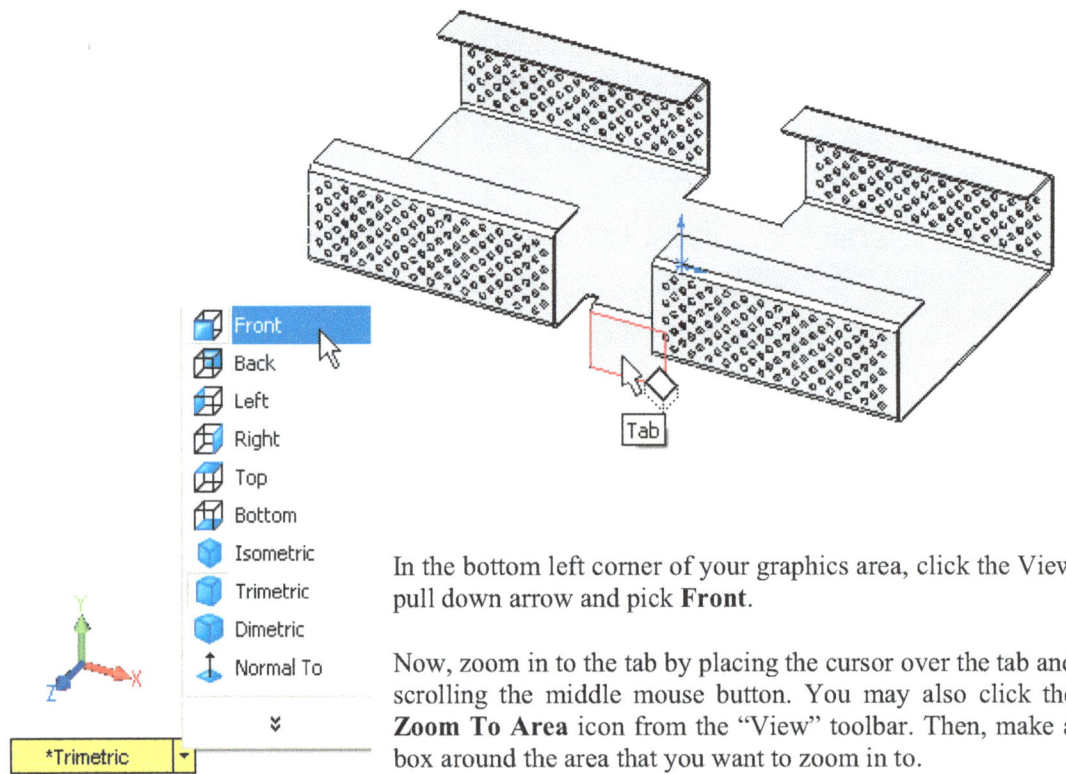

In the bottom left corner of your graphics area, click the View pull down arrow and pick **Front**.

Now, zoom in to the tab by placing the cursor over the tab and scrolling the middle mouse button. You may also click the **Zoom To Area** icon from the "View" toolbar. Then, make a box around the area that you want to zoom in to.

In the CommandManager, click the **Line** icon, or pull down the "Tools" menu and pick **Sketch Entities – Line**.

Create two parallel vertical lines.

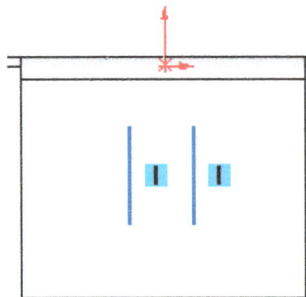

Then, click the **Tangent Arc** icon, or pull down the "Tools" menu and pick **Sketch Entities – Tangent Arc**.

Tangent Arc
Sketches an arc tangent to a sketch entity. Select the end point of a sketch entity, then drag to create the tangent arc.

Click on the top endpoint of the left vertical line. Move the cursor in the direction of the arc that you want and then click on the top endpoint of the right vertical line. Next, click on the bottom endpoint of the right vertical line. Move the cursor in the direction of the arc that you want and then click on the bottom endpoint of the left vertical line. Make sure that you move the cursor along the imaginary arc path. This ensures that the preview will be exactly what you want.

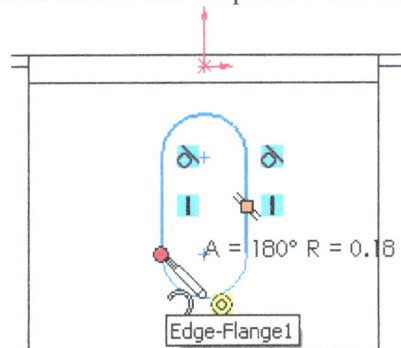

Press the **Escape** key on the keyboard to deselect the **Tangent Arc** tool.

Click on the center point of top arc. Hold down the **Ctrl** key and select the origin.

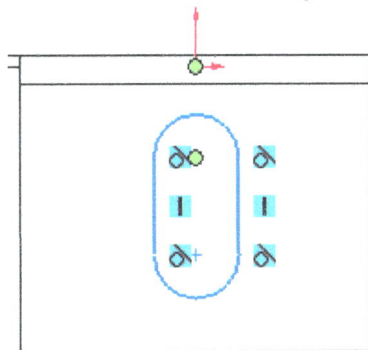

In the **Properties** PropertyManager, click the **Vertical** button.

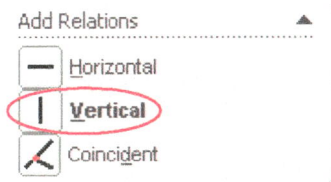

◈ Click the **Smart Dimension** icon in the CommandManager, or pull down the "Tools" menu and pick **Dimensions – Smart.**

Add a '**.125**' dimension to the bottom arc. Then, select both arcs and place a vertical dimension as shown. The value will be added in the next step.

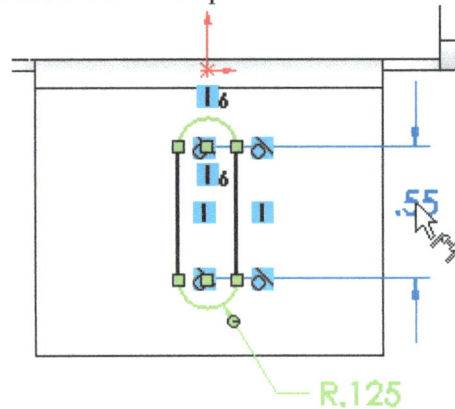

R.125

Right click on the vertical dimension and pick **Properties** from the menu.

In the **Dimension Properties** dialog box, change the **First arc condition** and the **Second arc condition** to **Max**.

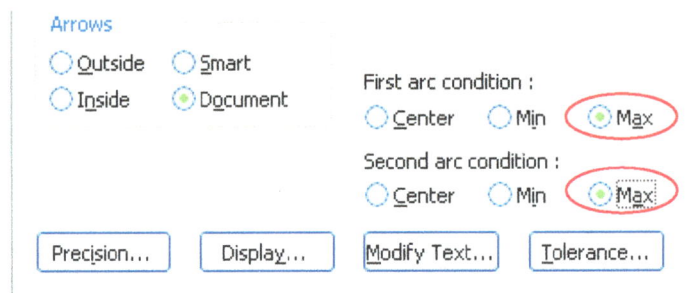

Click the **OK** button.

Double click on the vertical dimension and change it to '**.5**'.

⋮ Create a horizontal centerline connecting the midpoints of the vertical lines by clicking the **Centerline** icon in the CommandManager, or pull down the "Tools" menu and pick **Sketch Entities – Centerline**.

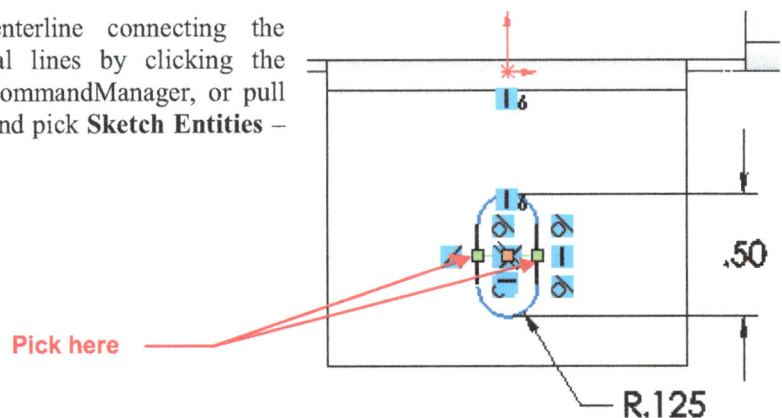

Pick here

.50

R.125

Click the **Smart Dimension** icon in the CommandManager, or pull down the "Tools" menu and pick **Dimensions – Smart.**

Click the bottom horizontal edge line of the tab and the centerline and place a '**.5**' dimension.

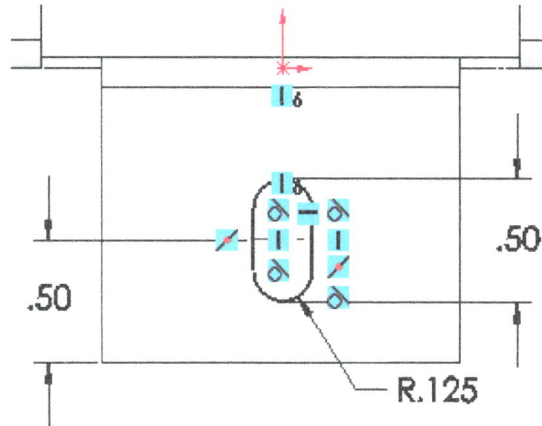

Exit the sketch by clicking the **Exit Sketch** icon in the CommandManager or in the upper right corner of the graphics area.

In the **Cut-Extrude** PropertyManager, check the **Link to thickness** check box and then click the green check mark button at the top of the **Cut-Extrude** PropertyManager.

Change back to the **Trimetric** view by clicking the View orientation pull down arrow in the bottom left corner of your graphics area and picking **Trimetric**.

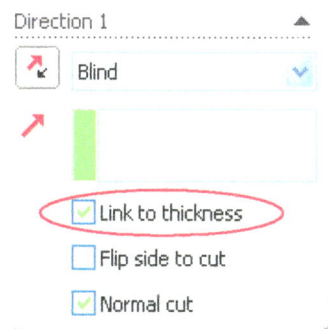

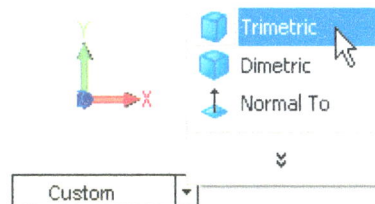

Rename the Feature

In the FeatureManager design tree, slowly click two times on **Cut-Extrude1** to select it.

Then, enter '**Obround**' for the new name.

Creating the Other Tab

Click the **Features** icon in the control area of the CommandManager. Then, click the **Mirror** icon from the toolbar, or pull down the "Insert" menu and pick **Pattern/Mirror – Mirror**.

Mirror
Mirrors features, faces, and bodies about a face or a plane.

In the upper left hand corner of the graphics area, click on the plus sign next to the document name (Part1) to expand the flyout FeatureManager design tree.

In the flyout FeatureManager design tree, select **Front Plane** as the **Mirror Face/Plane**.

Under **Features to Mirror**, select **Tab** and **Obround** from the flyout FeatureManager design tree.

Click the green check mark button at the top of the **Mirror** PropertyManager.

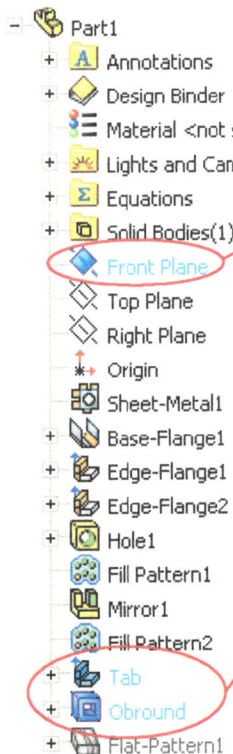

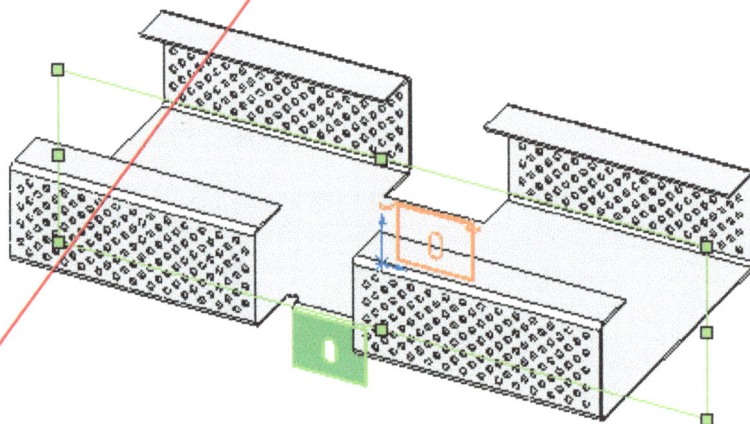

Mirror

Mirror Face/Plane
Front Plane

Features to Mirror
Tab
Obround

Faces to Mirror

Bodies to Mirror

Options
☐ Geometry Pattern
☑ Propagate Visual Properties

- 🔲 Part1
  - + 🅰 Annotations
  - + 🔷 Design Binder
  - 📄 Material <not specified>
  - + 🔦 Lights and Cameras
  - + Σ Equations
  - + 🔲 Solid Bodies(1)
  - 🔷 Front Plane
  - 🔷 Top Plane
  - 🔷 Right Plane
  - 🔺 Origin
  - 🔲 Sheet-Metal1
  - + 📐 Base-Flange1
  - + 📐 Edge-Flange1
  - + 📐 Edge-Flange2
  - + 🔲 Hole1
  - 🔲 Fill Pattern1
  - 🔲 Mirror1
  - 🔲 Fill Pattern2
  - + 📐 Tab
  - + 🔲 Obround
  - + 🔲 Flat-Pattern1

Cutout on Bottom

Click the View orientation pull down arrow in the bottom left corner of your graphics area and pick **Bottom**.

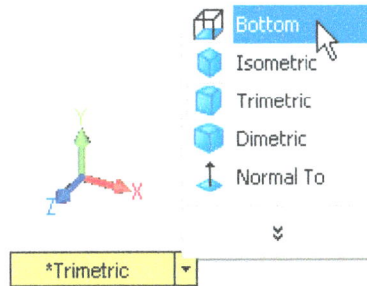

Click the **Sheet Metal** icon in the control area of the CommandManager. Then, click the **Extruded Cut** icon from the CommandManager or pull down the "Insert" menu and pick **Cut – Extrude**.

Select the bottom of the part for the plane to sketch on as shown.

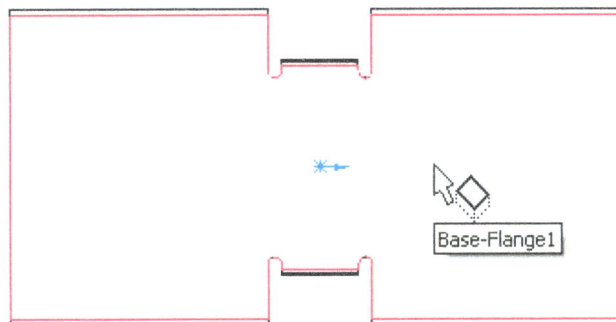

Create a rectangle on the left side of the part as shown on the next page using the **Rectangle** icon in the CommandManager, or pull down the "Tools" menu and pick **Sketch Entities – Rectangle**. You may need to zoom in to the left half of the part.

Create three lines on the top edge of the rectangle to create a cutout on the top of the rectangle using the **Line** icon in the CommandManager, or pull down the "Tools" menu and pick **Sketch Entities – Line**.

Right click in the graphics area and pick **End Chain** from the menu.

Create three lines on the bottom to create a cutout on the bottom of the rectangle.

Next, click the **Trim Entities** icon or pull down the "Tools" menu and pick **Sketch Tools - Trim Entities**.

Trim Entities
Trims or extends a sketch entity to be coincident to another, or deletes a sketch entity.

Trim to closest

In the **Trim** PropertyManager, make sure that the **Trim to closest** button is depressed, and select the middle section of the top and bottom horizontal lines to trim out the notch.

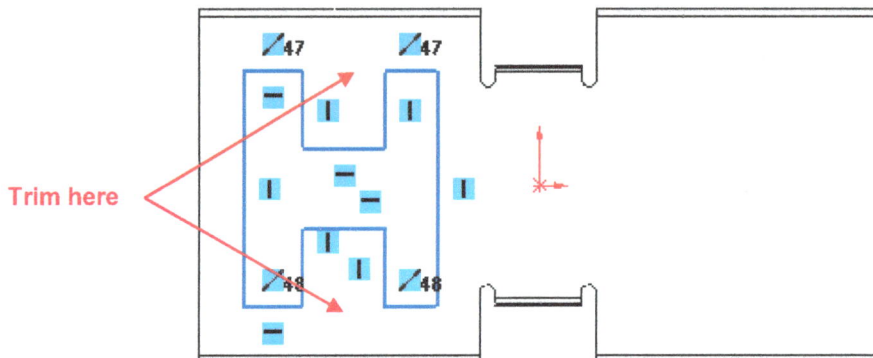

Trim here

Click the **Sketch Fillet** icon in the CommandManager, or pull down the "Tools" menu and pick **Sketch Tools – Fillet**.

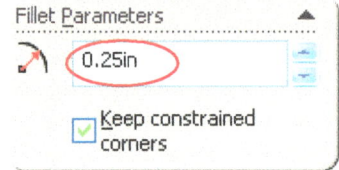

Fillet Parameters

0.25in

In the **Sketch Fillet** PropertyManager, set the **Radius** to '.25'.

Keep constrained corners

Select all the corners in the sketch.

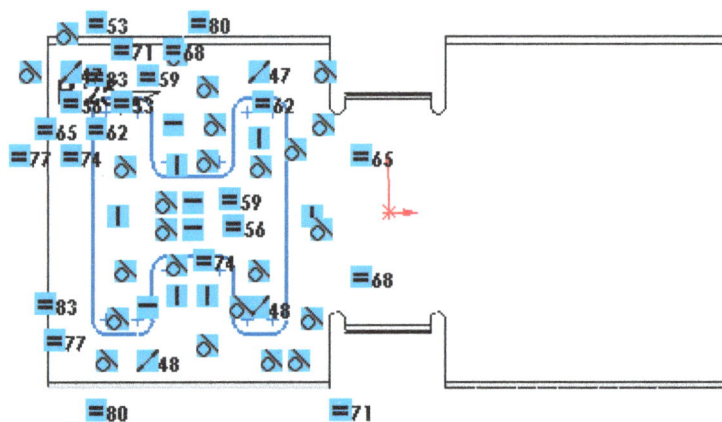

Click the green check mark button at the top of the **Sketch Fillet** PropertyManager.

By this time, the sketch relation flags are making the sketch very difficult to see. To hide these, pull down the "View" menu and pick **Sketch Relations**. If you need to see the sketch relation flags, simply pull down the "View" menu and pick **Sketch Relations** again.

Right click on left vertical line and pick **Select Midpoint** from menu. Hold down **Ctrl** key and select the origin.

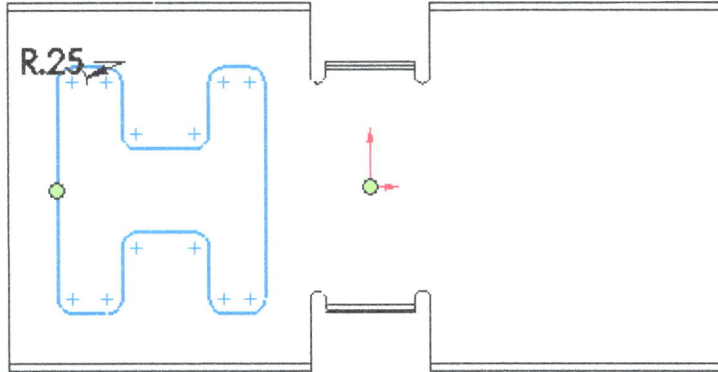

In the **Properties** PropertyManager, click the **Horizontal** button. ⬜ <u>H</u>orizontal

Select the upper left vertical line of the notch. Then, hold down **Ctrl** key and select the other three vertical lines as shown.

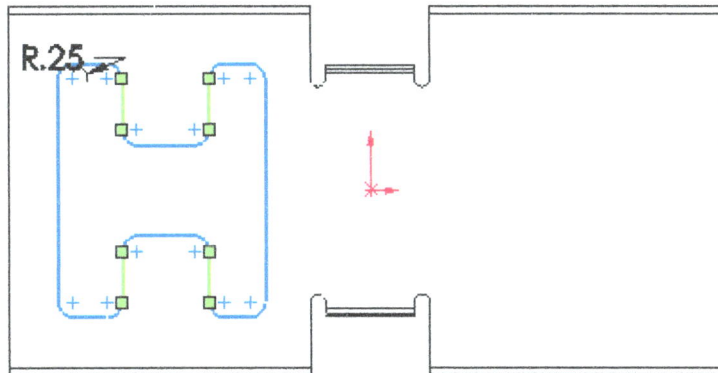

In the **Properties** PropertyManager, pick the **Equal** button. ⬜ E<u>q</u>ual

Left click and drag the dimension so that you can select the upper left horizontal line.

Select the upper left horizontal line. Then, hold down **Ctrl** key and select the other three horizontal lines as shown. If you are having trouble selecting the correct lines, zoom in closer.

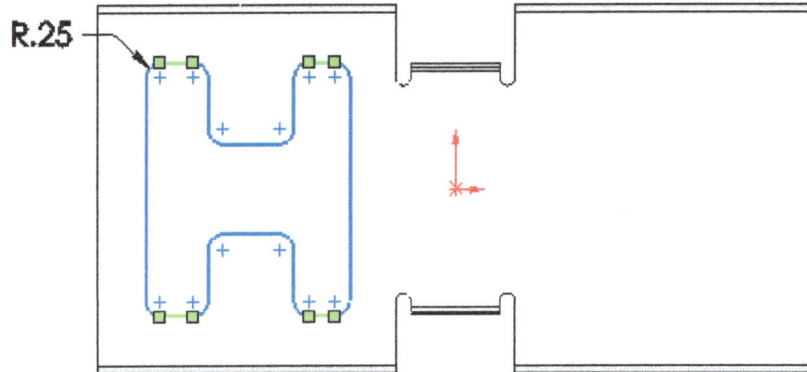

In the **Properties** PropertyManager, pick the **Equal** button. [= Equal]

Click the **Smart Dimension** icon in the CommandManager, or pull down the "Tools" menu and pick **Dimensions – Smart.**

Add a '**.5**' dimension and a '**.75**' dimension on the notch. Then, add a '**2**' dimension and a '**3**' dimension on the cutout. Add a '**1**' dimension from the side of the part to the left edge of the cutout as shown.

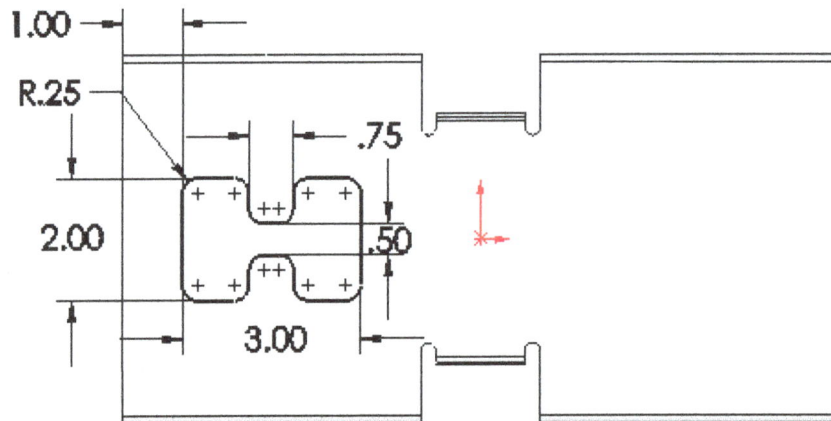

⊕ Create a circle in the upper tab by clicking the **Circle** icon in the CommandManager, or pull down the "Tools" menu and pick **Sketch Entities – Circle**.

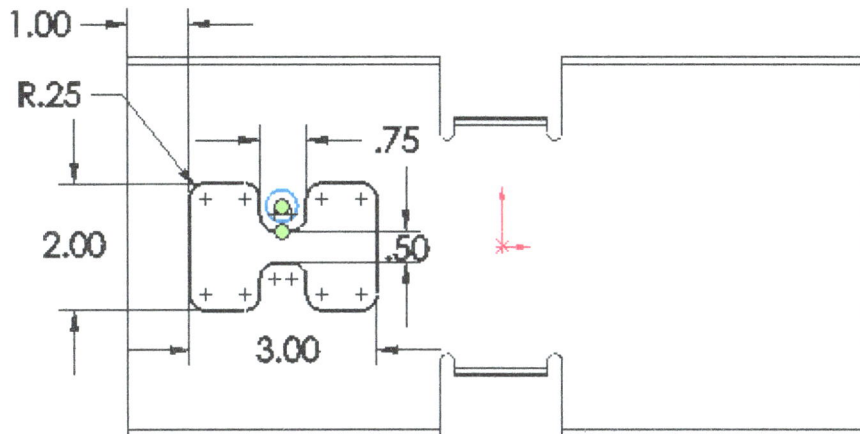

Press the **Escape** key on the keyboard to deselect the **Circle** tool.

Right click on the small horizontal line of the top notch and pick **Select Midpoint** from the menu. Make sure that the line is highlighted when you right click. Make sure that the line highlights and not the endpoint. You may need to Zoom in to select the line. Once the midpoint of the line is selected, hold down the **Ctrl** key and select the centerpoint of the circle.

In the **Properties** PropertyManager, pick the **Vertical** button. | | Vertical

✎ Click the **Smart Dimension** icon in the CommandManager, or pull down the "Tools" menu and pick **Dimensions – Smart**.

Add '**.25**' diameter dimension to the circle and a '**.3125**' dimension from the circle to the horizontal line.

⁝ Create a horizontal centerline through the origin by clicking the **Centerline** icon in the CommandManager, or pull down the "Tools" menu and pick **Sketch Entities – Centerline**.

Right click in the graphics area and pick **End Chain**. Create a vertical centerline through the origin.

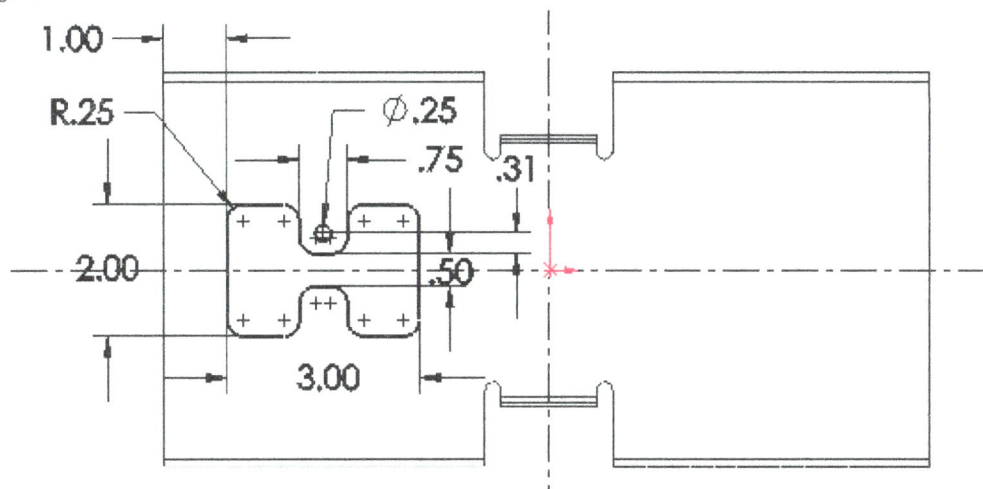

Press the **Escape** key on the keyboard to deselect the **Centerline** tool.

Click the **Mirror Entities** icon from the CommandManager, or pull down the "Tools" menu and pick **Sketch Tools – Mirror**.

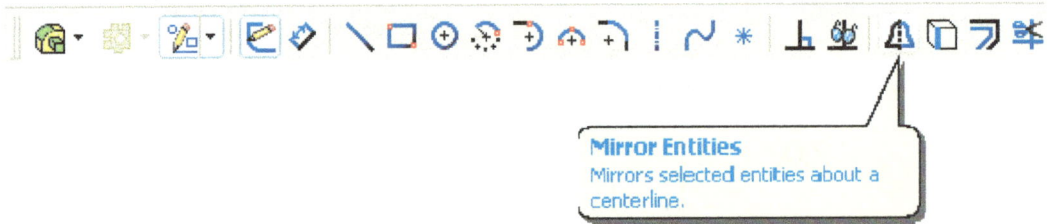

Mirror Entities
Mirrors selected entities about a centerline.

In the **Mirror** PropertyManager, click in the **Entities to mirror** box.

In the graphics area, select the circle.

Click in the **Mirror about** box and then select the horizontal centerline.

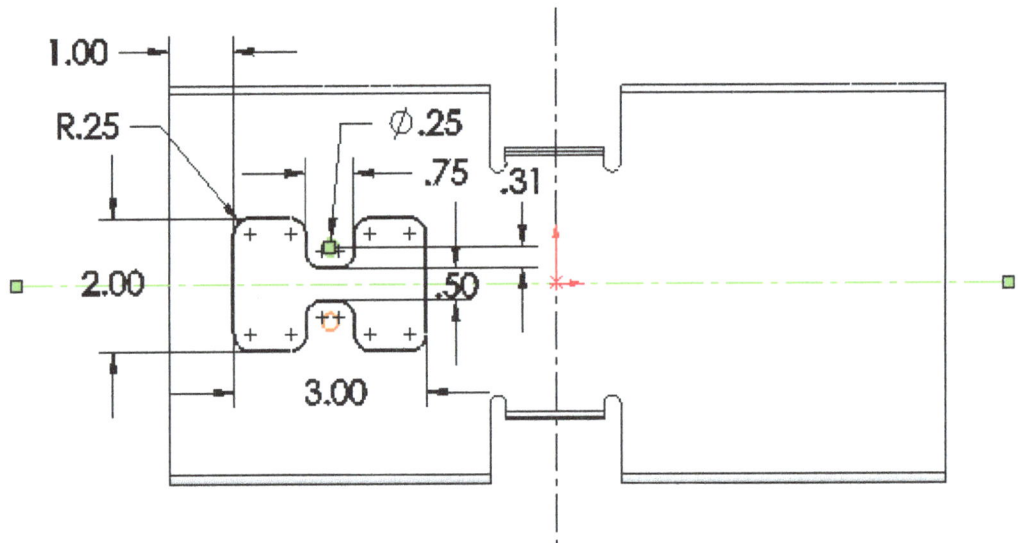

Click the green check mark button at the top of the **Mirror** PropertyManager.

Right click in the graphics area and pick **Recent Commands – Mirror Entities** from the menu.

In the **Mirror** PropertyManager, click in the **Entities to mirror** box.

In the graphics area, select all entities of the shape you created and the circles.

Click in the **Mirror about** box and select the vertical centerline.

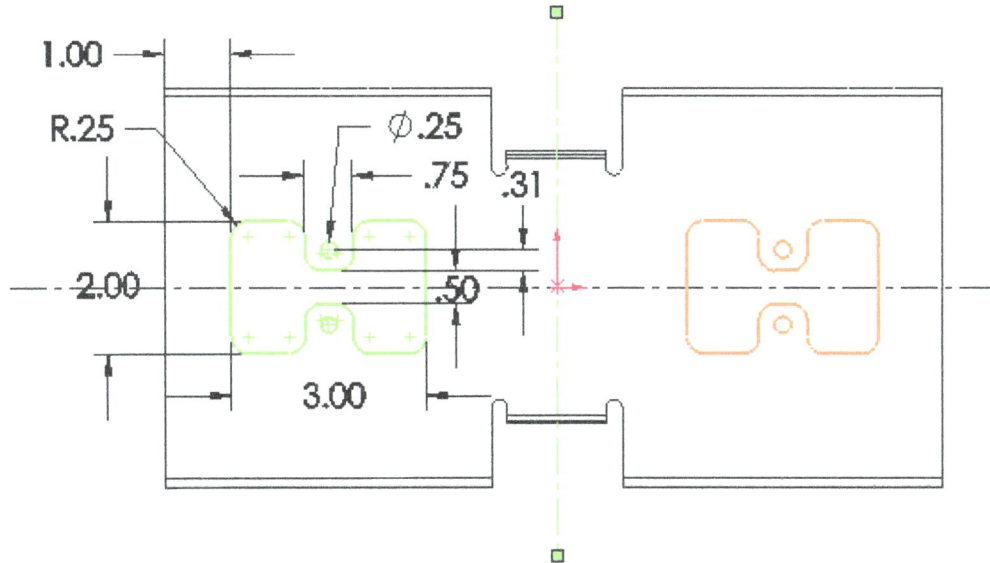

Click the green check mark button at the top of the **Mirror** PropertyManager.

Exit the sketch by clicking the **Exit Sketch** icon in the CommandManager or in the upper right corner of the graphics area.

In the **Cut-Extrude** PropertyManager, check the **Link to thickness** check box and then click the green check mark button at the top of the **Cut-Extrude** PropertyManager.

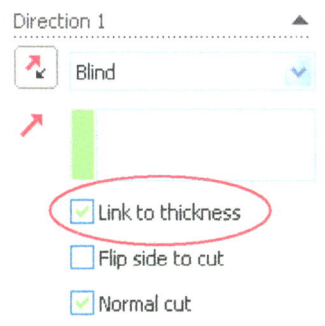

In the bottom left corner of your graphics area, click the pull down arrow and pick **Trimetric**.

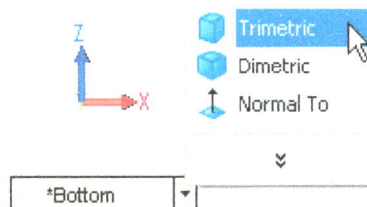

Save the Part

Click the **Save** icon in the "Standard" toolbar, or pick **Save** from the "File" pull down menu.

The **Save As** dialog box appears. In the **File name** box, type the name of the drawing number. For this chapter, use '**Housing Bracket**' and click **Save**.

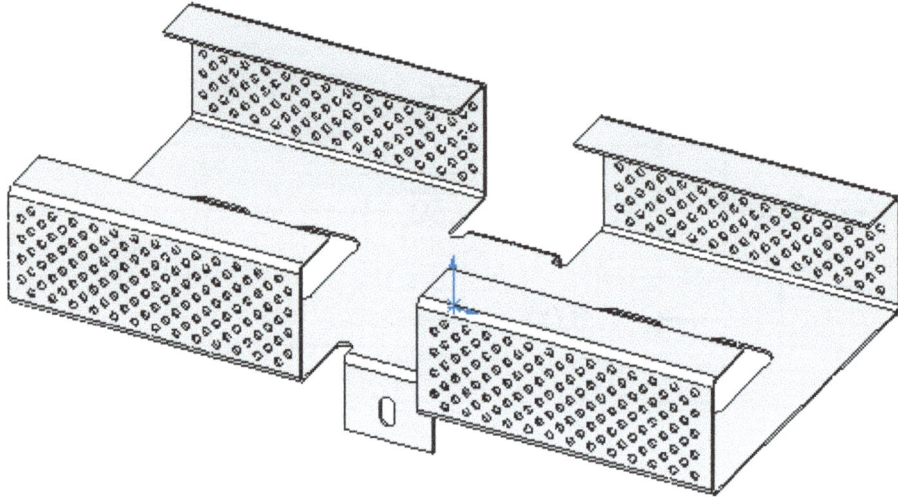

Rotate the part so that you can see all the features of the Housing Bracket that you created.

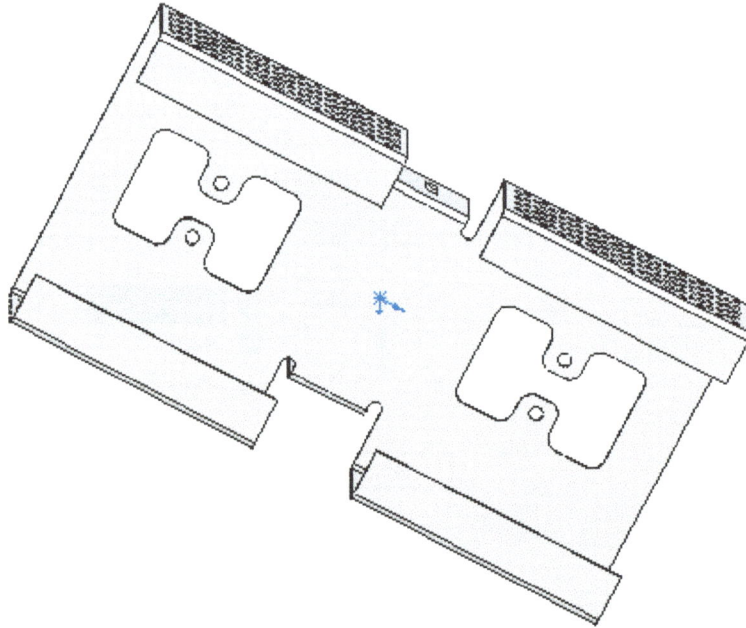

Chapter 3

Fingers

Fingers or multiple tabs are created by placing notches along the side of the part. A line pattern is used so that making adjustments to the size of the tab or the gap between the tabs can be easily modified.

The Jog command is a convenient method of creating the offset bend in the part. The Unfold and Fold features let you place the notch while in the flat so that the notch always passes through the entire bend area.

The slot on each finger is centered and the relationship to the tab ensures it always will be.

Again, a Mirror is used to simplify the part creation.

Create the Base Flange

Begin a new **Part** document by clicking the **New** icon in the "Standard" toolbar, or pull down the "File" menu and pick **New**.

Create a base flange by clicking the **Sheet Metal** icon in the control area of the CommandManager. Then, click the **Base-Flange/Tab** icon from the toolbar, or pull down the "Insert" menu and pick **Sheet Metal – Base Flange**.

Select the **Top** plane when prompted to select a plane on which to sketch the feature cross-section.

Create a rectangle with the lower left corner at the origin using the **Rectangle** icon in the CommandManager, or pull down the "Tools" menu and pick **Sketch Entities – Rectangle**.

Click the **Smart Dimension** icon in the CommandManager, or pull down the "Tools" menu and pick **Dimensions – Smart**.

Add a '**7**' vertical dimension to the left vertical line and a '**17.75**' horizontal dimension to the bottom horizontal line.

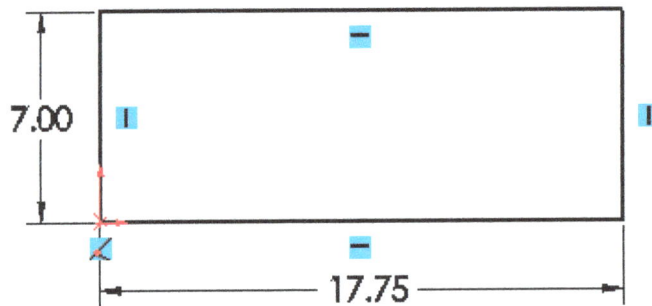

Exit the sketch by clicking the **Exit Sketch** icon in the CommandManager or in the upper right corner of the graphics area.

In the **Base Flange** PropertyManager under **Sheet Metal Parameters**, set the **Thickness** to '**.036**'. Make sure that the **Reverse direction** check box is checked.

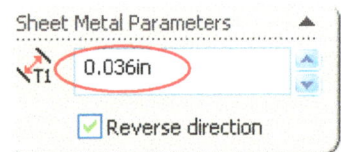

Click the green check mark button at the top of the **Base Flange** PropertyManager to accept the settings and create the part.

Create a Jog

Click the **Jog** icon in the CommandManager, or pull down the "Insert" menu and pick **Sheet Metal – Jog**.

Select the top of the part.

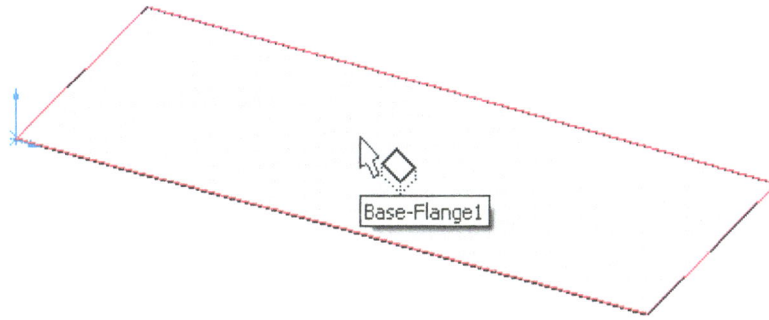

Create a horizontal line on the top of the part using the **Line** icon in the CommandManager, or pull down the "Tools" menu and pick **Sketch Entities – Line**.

Click the **Smart Dimension** icon in the CommandManager, or pull down the "Tools" menu and pick **Dimensions – Smart.**

Select the line and bottom edge of the part and add a '**4**' dimension as shown.

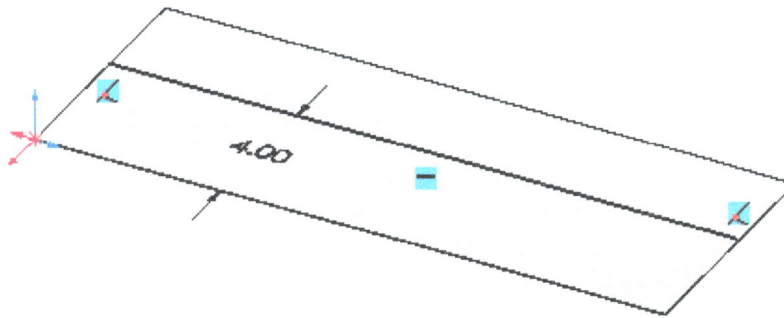

Exit the sketch by clicking the **Exit Sketch** icon in the CommandManager or in the upper right corner of the graphics area.

In the graphics area, select the top face of the part below the horizontal line. In the preview of the Jog feature, a black bullet appears where you selected, indicating the **Fixed Face**.

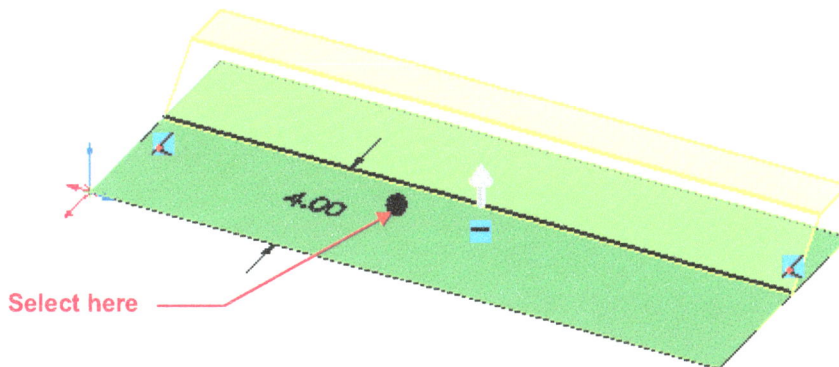

Select here

In the **Jog** PropertyManager, under **Jog Offset**, make sure that the **End Condition** is set to **Blind** and enter '**1.5**' for the **Offset Distance**.

Click the **Overall Dimension** button and the **Material Inside** button. Set the **Jog Angle** set to '**45**'.

Check the green check mark button at the top of the **Jog** PropertyManager to accept the settings and create the feature.

Unfold the Jog

In order to create the proper cut for the fingers, you must first unbend the previous jog feature.

Click the **Unfold** icon in the CommandManager, or pull down the "Insert" menu and pick **Sheet Metal – Unfold**.

For the **Fixed face**, select the bottom flat face of the part.

For the **Bends to unfold**, select the two bend areas from the edge flanges.

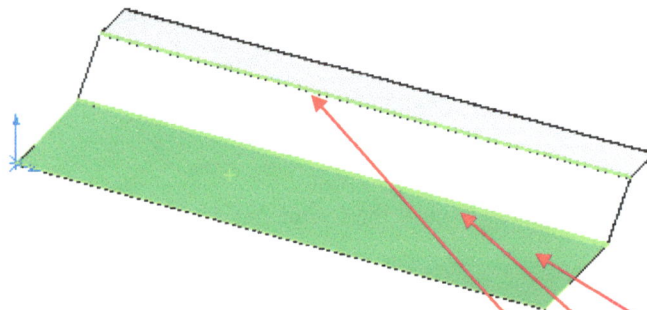

Click the green check mark button at the top of the **Unfold** PropertyManager and the bends will unfold.

Create the Fingers

Click the **Extruded Cut** icon from the CommandManager or pull down the "Insert" menu and pick **Cut – Extrude**.

Extruded Cut
Cuts a solid model by extruding a sketched profile in one or two directions.

Select the top of the part for the plane to sketch on as shown.

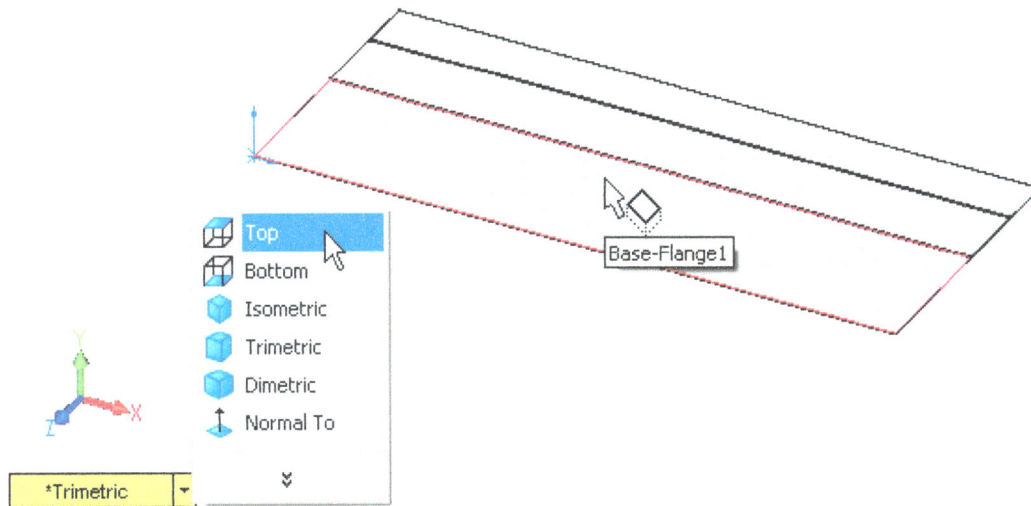

Base-Flange1

Top
Bottom
Isometric
Trimetric
Dimetric
Normal To

*Trimetric

In the bottom left corner of the graphics area, change the View orientation by clicking the pull down arrow and picking **Top**, or press **Ctrl-5**.

Create a rectangle with the lower left corner on the bottom bend line as shown using the **Rectangle** icon in the CommandManager, or pull down the "Tools" menu and pick **Sketch Entities – Rectangle**.

Click the **Smart Dimension** icon in the CommandManager, or pull down the "Tools" menu and pick **Dimensions – Smart**.

Add a '1' vertical dimension from the top of the part to the top horizontal line of the rectangle. Add a '.25' horizontal dimension to the two vertical lines of the rectangle. Then, add a '1.25' horizontal dimension from the left edge of the part to the left vertical line of the rectangle.

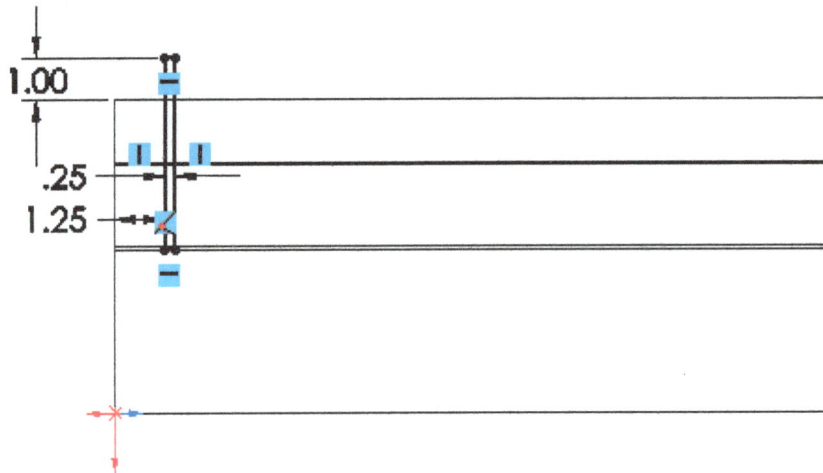

Pull down the "Tools" menu and pick **Sketch Tools – Linear Pattern**.

In the graphics area, select the four lines of the rectangle for the **Entities to Pattern**. You may need to zoom in closer to select the small horizontal lines.

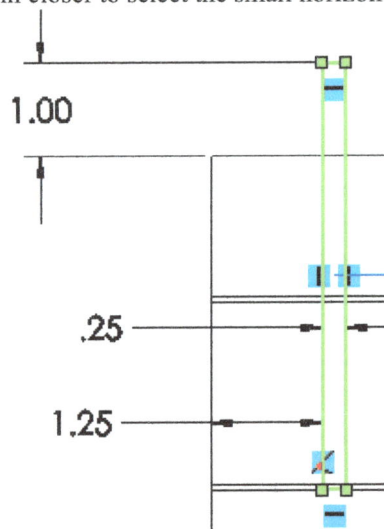

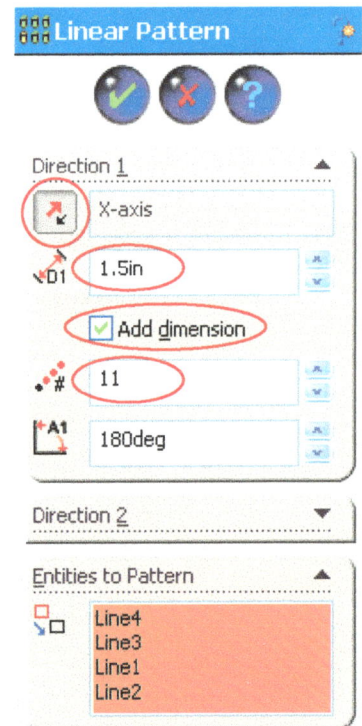

In the **Linear Pattern** PropertyManager, set the spacing to '1.5'. Make sure the **Add dimension** check box is checked to display a dimension between entities.

Set the **Number** to '11'.

If the preview of the linear pattern is off the part, change the direction of the pattern by clicking the **Reverse Direction** button.

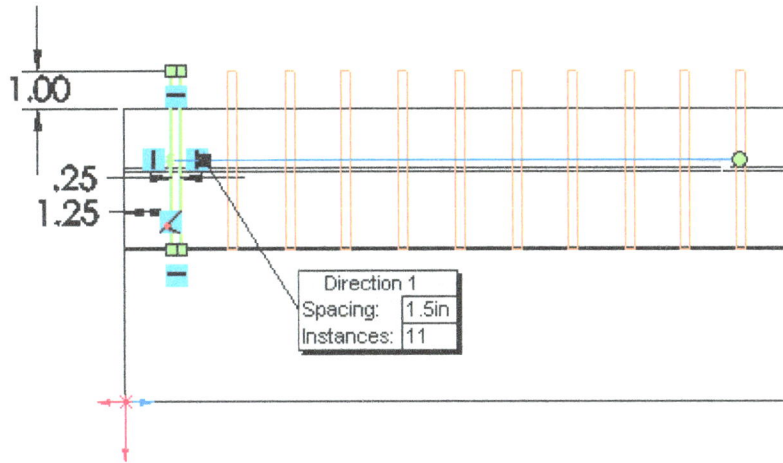

Click the green check mark button at the top of the **Linear Pattern** PropertyManager.

Exit the sketch by clicking the **Exit Sketch** icon in the CommandManager or in the upper right corner of the graphics area.

In the **Cut-Extrude** PropertyManager, check the **Link to thickness** check box and then click the green check mark button at the top of the **Cut-Extrude** PropertyManager.

Change back to the **Trimetric** view by clicking the View orientation pull down arrow in the bottom left corner of your graphics area and picking **Trimetric**.

Fold the Jog

Click the **Sheet Metal** icon in the control area of the CommandManager. Then, click the **Fold** icon in the CommandManager, or pull down the "Insert" menu and pick **Sheet Metal – Fold**.

In the **Fold** PropertyManager, click on the **Collect All Bends** button to select all the appropriate bends in the part.

Click the green check mark button at the top of the **Fold** PropertyManager to fold the part back up.

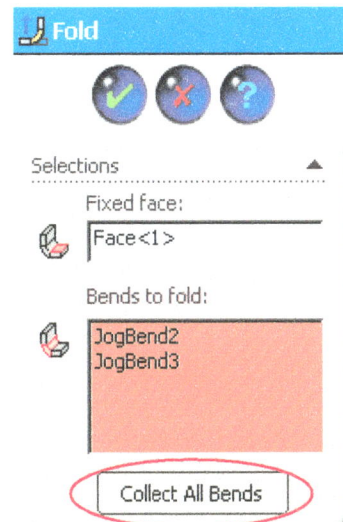

Add a Cut in the Finger

Click the **Extruded Cut** icon from the CommandManager or pull down the "Insert" menu and pick **Cut – Extrude**.

Extruded Cut
Cuts a solid model by extruding a sketched profile in one or two directions.

Select the top of the leftmost finger for the plane to sketch on as shown.

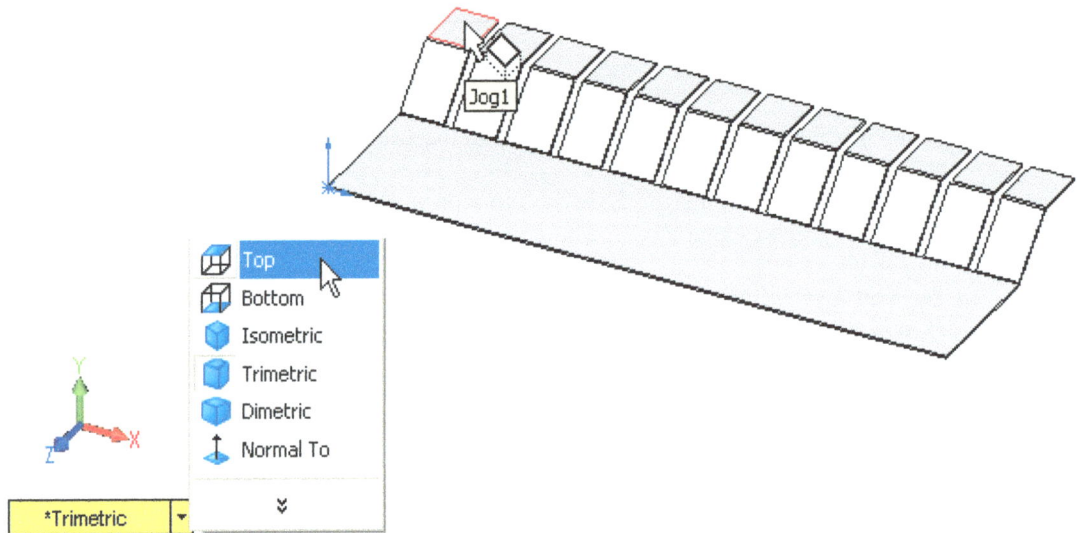

In the bottom left corner of the graphics area, change the View orientation by clicking the pull down arrow and picking **Top**, or press **Ctrl-5**.

Create a rectangle in the middle of the finger using the **Rectangle** icon in the CommandManager, or pull down the "Tools" menu and pick **Sketch Entities – Rectangle**.

Then, click the **Tangent Arc** icon, or pull down the "Tools" menu and pick **Sketch Entities – Tangent Arc**.

Click on the bottom right endpoint of the rectangle. Move the cursor the direction of the arc that you want and then click on the bottom left endpoint of the rectangle.

Press the **Escape** key on the keyboard to deselect the **Tangent Arc** tool.

Select the bottom horizontal line of the rectangle and press the **Delete** key on the keyboard.

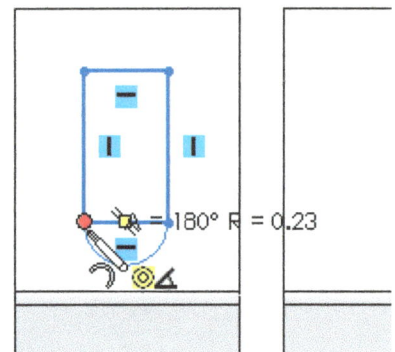

Right click on the top horizontal line of the rectangle and pick **Select Midpoint** from the menu. Hold down the **Ctrl** key and select the top edge line of the part as shown.

In the **Properties** PropertyManager, click the **Midpoint** button.

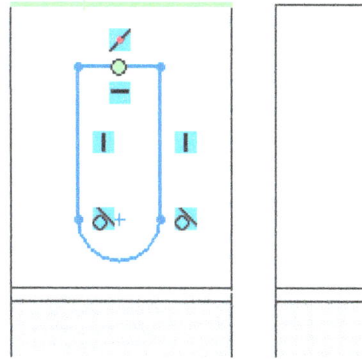

Click the **Smart Dimension** icon in the CommandManager, or pull down the "Tools" menu and pick **Dimensions – Smart**.

Add a '**.75**' vertical dimension to the left vertical line. Then, add a '**9/32**' horizontal dimension to the two vertical lines of the rectangle. In the **Modify** dialog box, you can type in a fractional value and SolidWorks will calculate the decimal value for you.

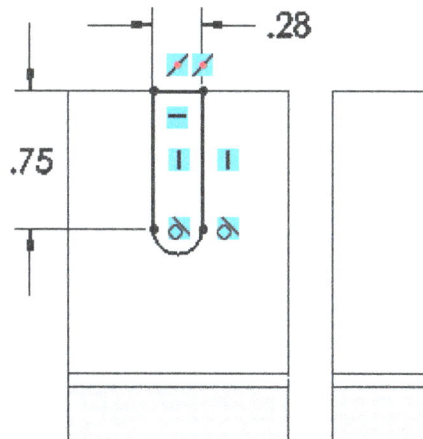

Exit the sketch by clicking the **Exit Sketch** icon in the CommandManager or in the upper right corner of the graphics area.

In the **Cut-Extrude** PropertyManager, check the **Link to thickness** check box and then click the green check mark button at the top of the **Cut-Extrude** PropertyManager.

Change back to the **Trimetric** view by clicking the View orientation pull down arrow in the bottom left corner of your graphics area and picking **Trimetric**.

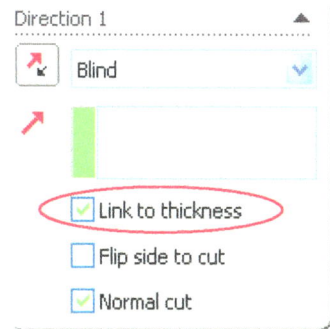

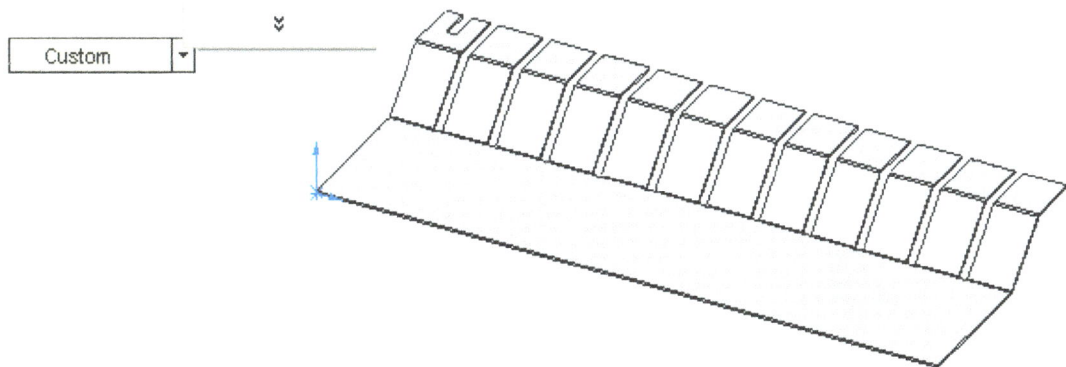

Copy the Cutout

Click the **Features** icon in the control area of the CommandManager. Then, click the **Linear Pattern** icon in the CommandManager, or pull down the "Insert" menu and pick **Pattern/Mirror – Linear Pattern**.

Linear Pattern
Patterns features, faces, and bodies in one or two linear directions.

Select the front edge of the part near the left end as shown to set the pattern direction. After you select the edge, you should see a large grey arrow on the origin pointing to the right, the positive X-direction.

In the **Linear Pattern** PropertyManager, set the **Spacing** to '**1.5**'.

Set the **Number** to '**12**'.

Click in the **Features to Pattern** box. In the flyout FeatureManager design tree, select **Cut-Extrude2**. This may already be selected for you.

Click the green check mark button at the top of the **Linear Pattern** PropertyManager.

Mirror the Part

Click the **Mirror** icon in the CommandManager, or pull down the "Insert" menu and pick **Pattern/Mirror – Mirror**.

Mirror
Mirrors features, faces, and bodies about a face or a plane.

In the graphics area, select the small surface of the front of the part on the midline. You may need to zoom in to see select it.

In the **Mirror** PropertyManager, click in the **Bodies to Mirror** box.

In the graphics area, select the part as shown.

Click the green check mark button at the top of the **Mirror** PropertyManager.

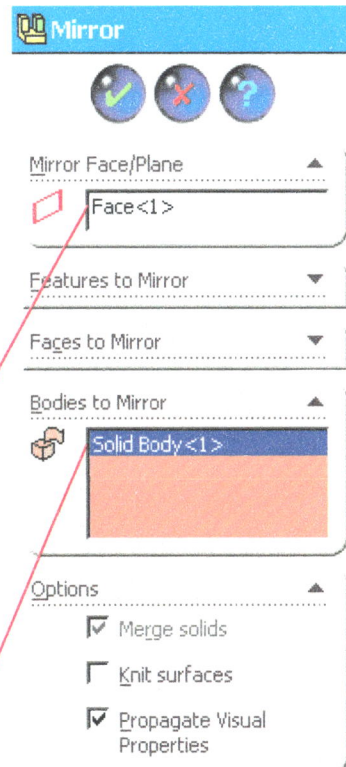

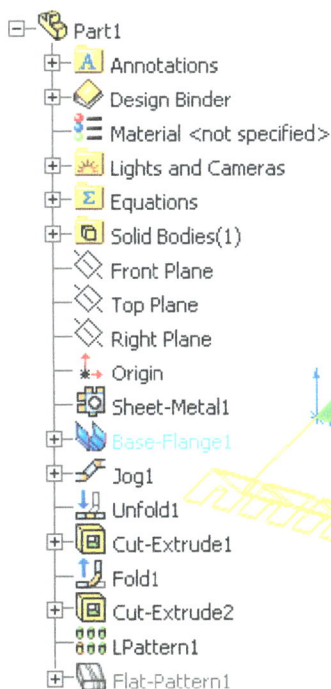

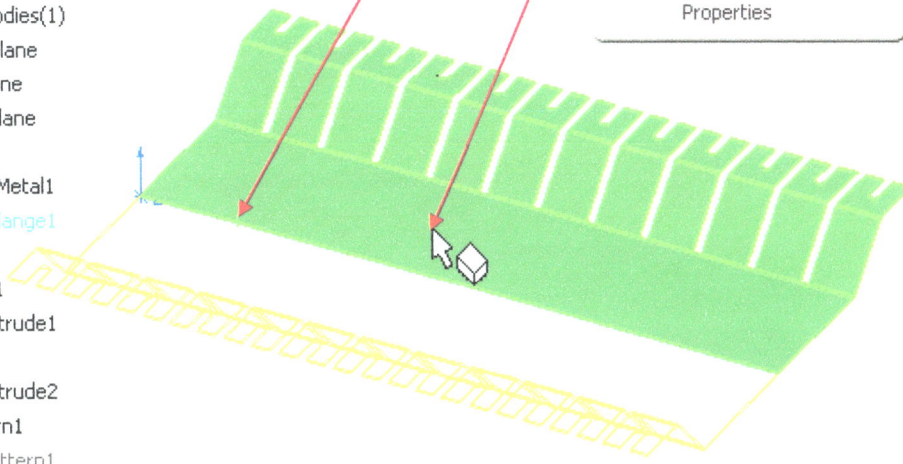

Mirror

Mirror Face/Plane
Face<1>

Features to Mirror

Faces to Mirror

Bodies to Mirror
Solid Body<1>

Options
☑ Merge solids
☐ Knit surfaces
☑ Propagate Visual Properties

Part1
  Annotations
  Design Binder
  Material <not specified>
  Lights and Cameras
  Equations
  Solid Bodies(1)
  Front Plane
  Top Plane
  Right Plane
  Origin
  Sheet-Metal1
  Base-Flange1
  Jog1
  Unfold1
  Cut-Extrude1
  Fold1
  Cut-Extrude2
  LPattern1
  Flat-Pattern1

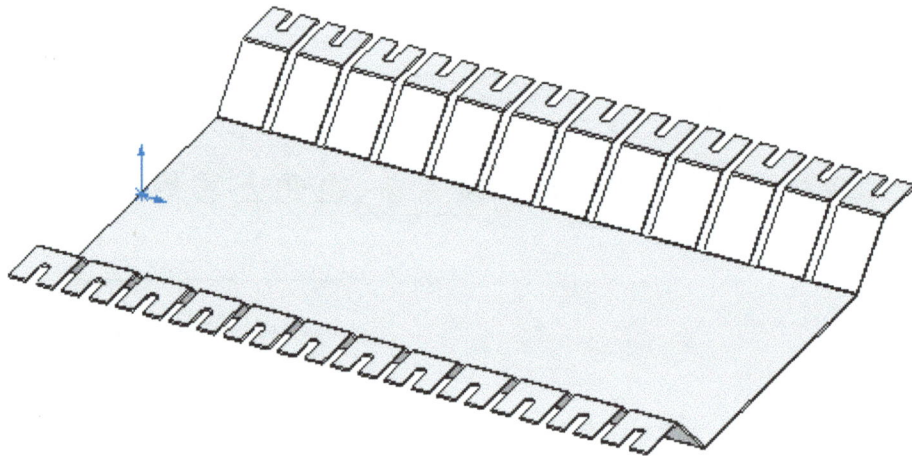

Save the Part

Click the **Save** icon in the "Standard" toolbar, or pick **Save** from the "File" pull down menu.

The **Save As** dialog box appears.

In the **File name** box, type the name of the drawing number. For this chapter, use '**Fingers**' and select **Save**.

Closing the File

Pick **Close** from the "File" pull down menu.

Click **No** when prompted to **Save changes to Fingers.SLDPRT?**

Chapter 4

LED Cover Plate

The LED Cover Plate utilizes a Sketch Driven Pattern to create an exit sign. The Display Grid is used during sketching to make the layout of the points easier.

The Hole Wizard is a good tool to use when you have a quantity of holes of the same size. Another nice feature of the Hole Wizard is that it appropriately renames the feature in the Feature Manager design tree for you. If you later decide to change the hole size, the Hole Wizard magically changes the feature name as well.

Create the Base Flange

Begin a new **Part** document by clicking the **New** icon in the "Standard" toolbar, or pull down the "File" menu and pick **New**.

Create a base flange by clicking the **Sheet Metal** icon in the control area of the CommandManager. Then, click the **Base-Flange/Tab** icon from the toolbar, or pull down the "Insert" menu and pick **Sheet Metal – Base Flange**.

Select the **Front** plane when prompted to select a plane on which to sketch the feature cross-section.

Create a rectangle with the lower left corner at the origin using the **Rectangle** icon in the CommandManager, or pull down the "Tools" menu and pick **Sketch Entities – Rectangle**.

Click the **Smart Dimension** icon in the CommandManager, or pull down the "Tools" menu and pick **Dimensions – Smart**.

Add a '**4**' vertical dimension to the left vertical line and a '**6**' horizontal dimension to the bottom horizontal line.

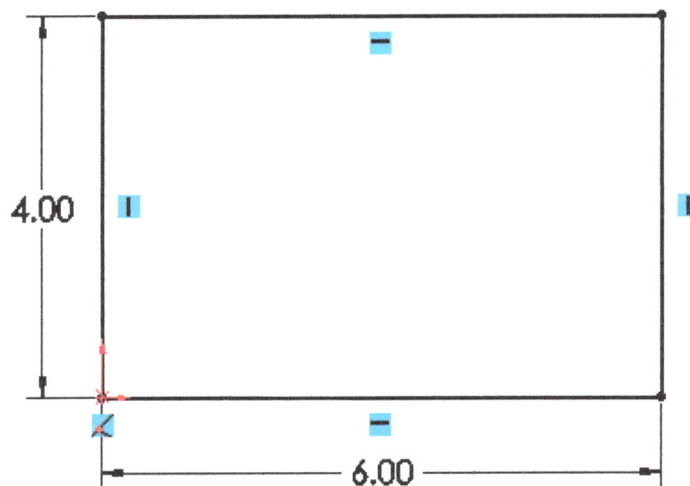

You can display a sketch grid in the active sketch and set options for the grid display and snap functionality. To set these options, pull down the "Tools" menu and pick **Options**.

In the **System Options** dialog box, under **Sketch**, click on **Relations/Snaps**.

Check **Enable snapping** at the top of the dialog box. Also, check **Grid** near the bottom of the list and make sure that **Snap only when grid is displayed** is checked.

Then, click on the **Go To Document Grid Settings** button at the top right of the dialog box.

System Options | Document Properties |

- General
- Drawings
  - Display Style
  - Area Hatch/Fill
- Colors
- Sketch
  - Relations/Snaps
- Display/Selection
- Performance
- Assemblies
- External References
- Default Templates
- File Locations
- FeatureManager
- Spin Box Increments
- View Rotation
- Backups
- Data Options
- File Explorer
- Collaboration

☑ Enable snapping   Go To Document Grid Settings

☑ Snap to model geometry

☑ Automatic relations

Sketch Snaps
☑ End points and sketch points
☑ Center Points
☑ Mid-points
☑ Quadrant Points
☑ Intersections
☑ Nearest
☑ Tangent
☑ Perpendicular
☑ Parallel
☑ Horizontal/vertical lines
☑ Horizontal/vertical to points
☑ Length
☑ Grid
    ☑ Snap only when grid is displayed
☑ Angle
    Snap angle: [45.00deg ⇅]

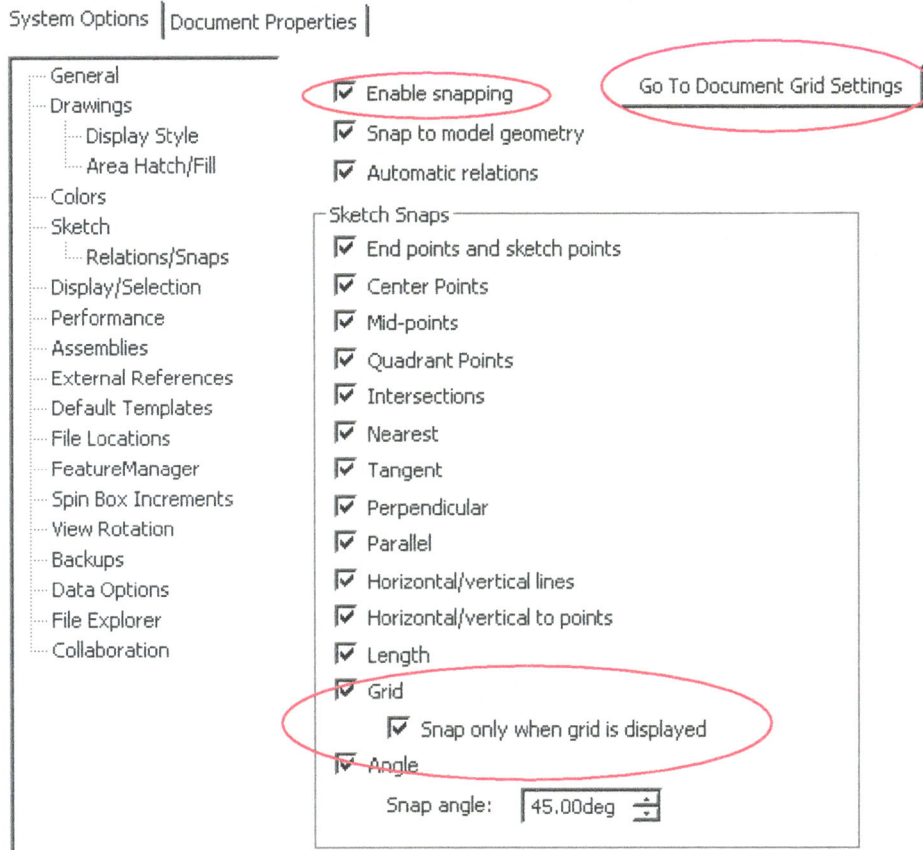

Under **Document Properties**, check **Display grid** and **Automatic scaling**. Uncheck the **Dash** check box. Set the **Major grid spacing** to '.5', the **Minor-lines per major** to '4', and the **Snap points per minor** to '1'.

System Options | Document Properties |

- Detailing
  - Dimensions
  - Notes
  - Balloons
  - Arrows
  - Virtual Sharps
  - Annotations Display
  - Annotations Font
- Grid/Snap
- Units
- Colors
- Material Properties
- Image Quality
- Plane Display

Grid
☑ Display grid
☐ Dash
☑ Automatic scaling

Major grid spacing: [0.50in ⇅]

Minor-lines per major: [4 ⇅]

Snap points per minor: [1 ⇅]

Go To System Snaps

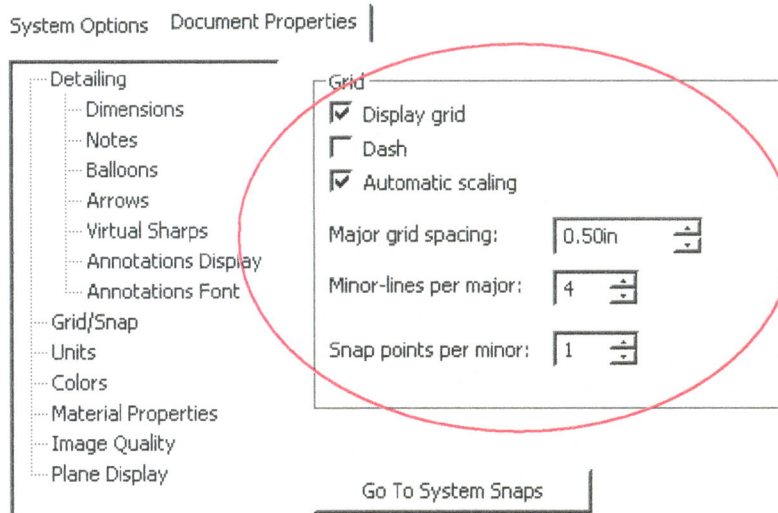

Click the **OK** button to accept the setting changes and close the dialog box.

* Click on the **Point** icon in the CommandManager, or pull down the "Tools" menu and pick **Sketch Entities – Point**.

You may need to **Zoom to Fit** to see the entire sketch. In the sketch, create points on the grid points as shown to spell out the word 'EXIT'. You may need to zoom in closer to see more grid lines. A small preview of the point appears when you snap to each grid point.

If you create a point that you did not want, try **Undo**. Otherwise, press the **Escape** key, select the point and press the **Delete** key. Then, select the **Point** icon and continue to create the points.

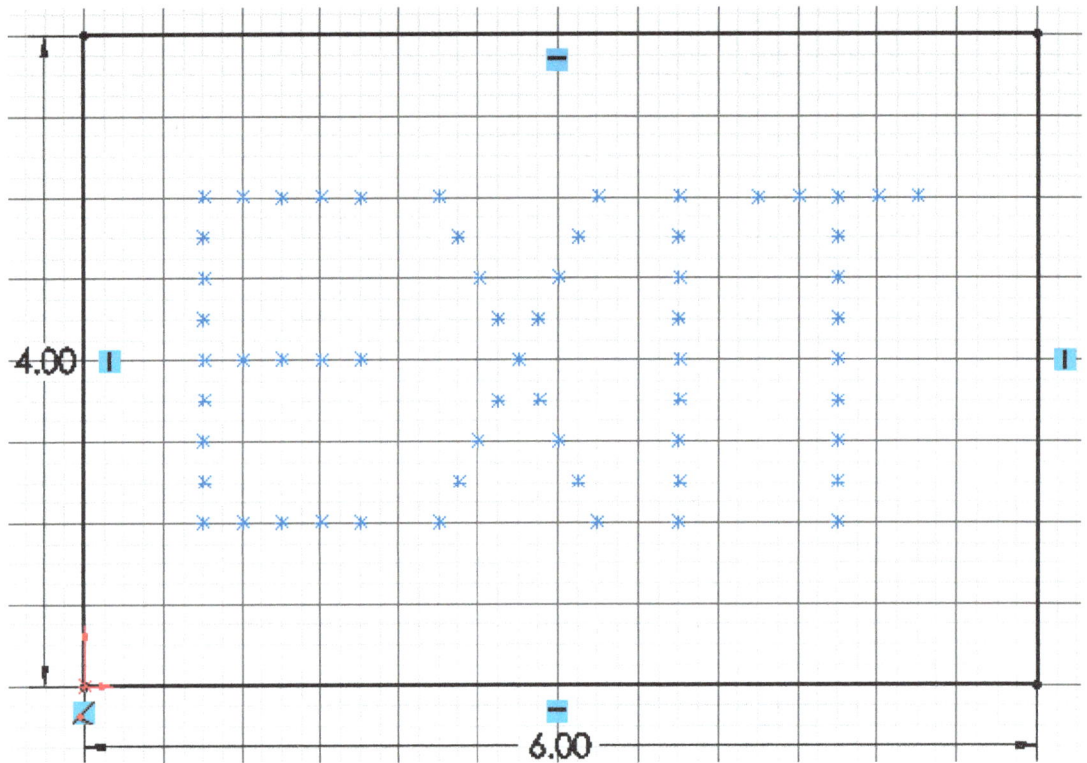

Exit the sketch by clicking the **Exit Sketch** icon in the CommandManager or in the upper right corner of the graphics area.

In the **Base Flange** PropertyManager under **Sheet Metal Parameters**, set the **Thickness** to '1/8'. Make sure that the **Reverse direction** check box is not checked.

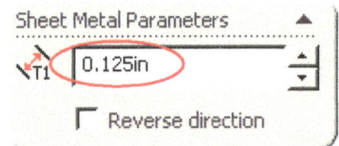

Click the green check mark button at the top of the **Base Flange** PropertyManager to accept the settings and create the part.

In the Feature Manager design tree, click on the plus sign next to **Base-Flange1** to expand the feature.

Right click on **Sketch1** and pick **Show** from the menu.

Hole Wizard

Click the **Features** icon in the control area of the CommandManager. Then, click the **Hole Wizard** icon from the toolbar, or pull down the "Insert" menu and pick **Features – Hole – Wizard**.

Hole Wizard
Inserts a hole using a pre-defined cross-section.

In the **Hole Specification** PropertyManager, under the **Type** tab, click the **Hole** button.

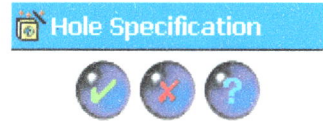

Change the **Standard** to **Ansi Metric** and the **Type** to **Screw Clearances**. Set the **Size** to **M3** and the **End Condition** to **Up to Next**.

Then, click the **Positions** tab, and select a position below the EXIT sketch as shown.

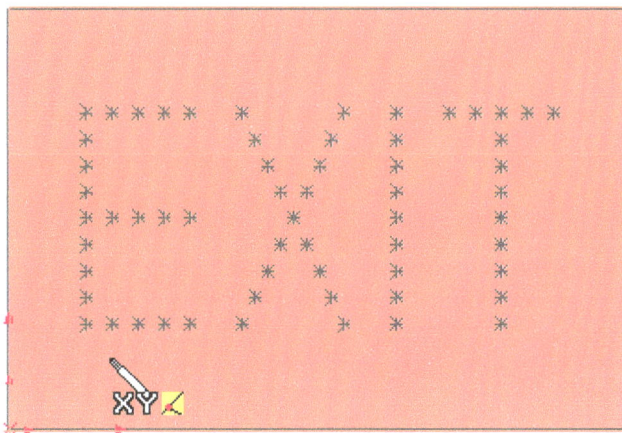

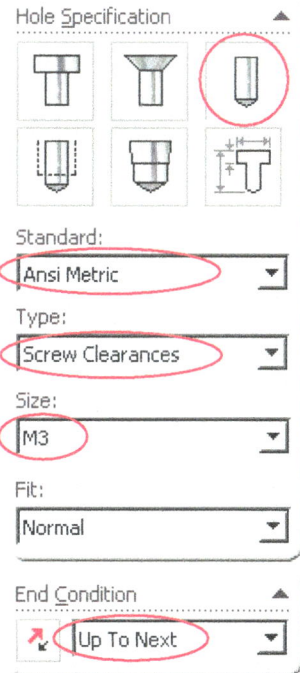

Press the **Escape** key to deselect the **Point** tool.

Drag the point to the lower left point on the EXIT sketch as shown.

Click the green check mark button at the top of the **Point** PropertyManager, and then, click the green check mark button at the top of the **Hole Specification** PropertyManager.

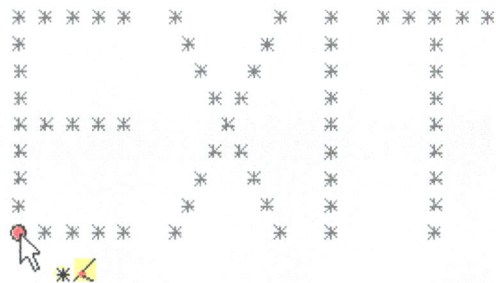

Sketch Driven Pattern

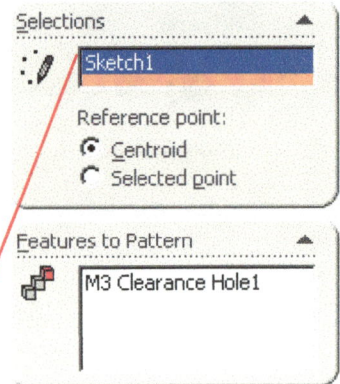

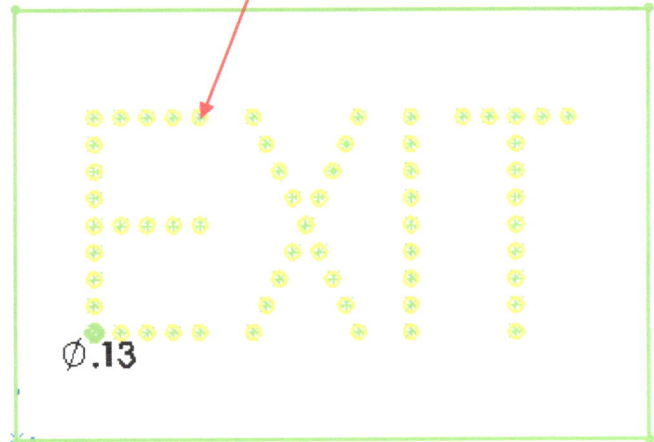

Pull down the "Insert" menu and pick **Pattern/Mirror – Sketch Driven Pattern**.

In the **Sketch Driven Pattern** PropertyManager, click in the **Selections** box. Then, in the graphics area, select any of the points in order to select **Sketch1**.

If **M3 Clearance Hole1** is not already selected for you under **Features to Pattern**, click in the **Features to Pattern** box. In the flyout FeatureManager design tree, select **M3 Clearance Hole1**.

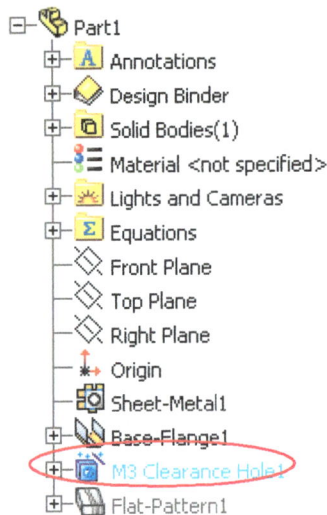

Click the green check mark button at the top of the **Sketch Driven Pattern** PropertyManager.

In the Feature Manger design tree, right click on **Sketch1** and pick **Hide** from the menu.

Click the View orientation pull down arrow in the bottom left corner of your graphics area and pick **Trimetric**.

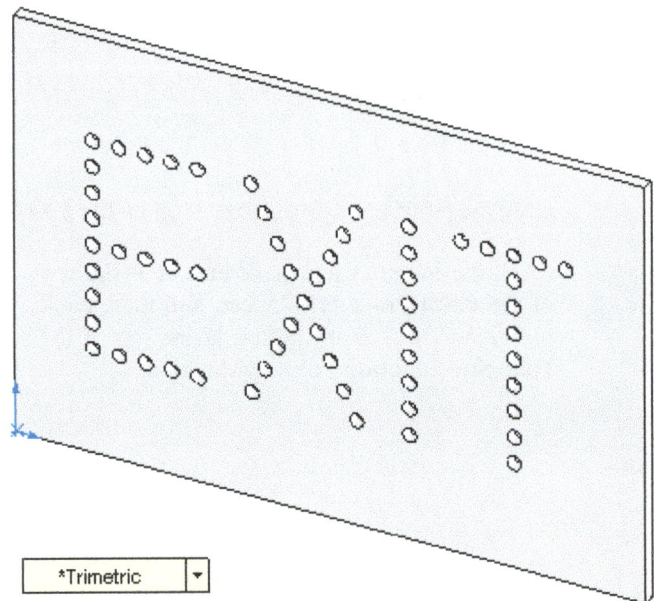

Saving the Part

Click the **Save** icon in the "Standard" toolbar, or pick **Save** from the "File" pull down menu. In the **File name** box, type '**LED Cover Plate**' and select **Save**.

Chapter 5

Design Library

The Design Library is a great concept. However, the tools you want to use are scattered throughout it. In this chapter, you will create your own folder in the Design Library and start to place your favorite tools in it.

Subfolders are used to separate hole shapes from formed shapes. A special attribute must be set on the form tool folder. Form tools are actually the dies which are used to form or bend the material into the desired shape. Forming tools can only be used from the Design Library and only apply to sheet metal parts.

Commonly used features, such as holes or slots, can be created and saved as library features. This provides a great time savings and also makes the geometry consistent among your CAD files. Think about selecting the desired obround from a list rather than drawing it new each time you need one.

The Design Library

The Task Pane appears on the right hand side of your screen when you open SolidWorks.

If the Task Pane is hidden, pull down the "View" menu and pick **Task Pane**.

With the Task Pane shown, click the **Design Library** icon in the Task Pane. To expand or collapse the Task Pane, click an arrow or anywhere along the bar between the arrows.

> **Design Library**
> Click to display this task pane tab.

The **Design Library** tab in the Task Pane provides a central location for reusable elements such as features, parts, and even forming tools.

In the **Design Library** tab, click the **Add File Location** icon.

In the **Choose Folder** dialog box, browse to the default installation folder of SolidWorks, usually on your local hard drive in the Program Files folder. Then, double click on the **data** folder.

While in the **data** folder, in the **Choose Folder** dialog box, click the **Create New Folder** icon.

A new folder is created in the **data** folder. Enter 'SheetMetalGuy' for the folder name.

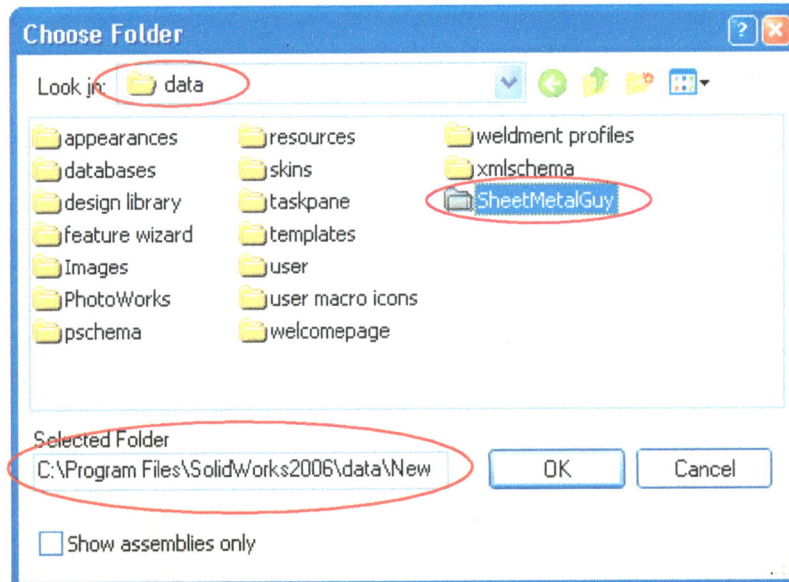

Double click on the **SheetMetalGuy** folder and then click the **OK** button.

The **SheetMetalGuy** Design Library has now been created.

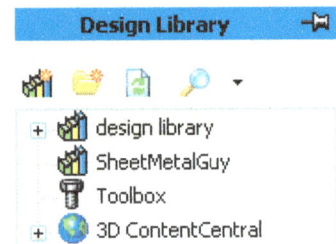

✳ In the **Design Library** tab, right click on **SheetMetalGuy** and pick **New Folder** from the menu. You may also click on **SheetMetalGuy** to highlight it and then click on the **Create New Folder** icon.

A new folder is created under **SheetMetalGuy**. Enter '**Punching Shapes**' for the folder name.

Copy an Existing Library Feature

A library feature is a frequently used feature, or combination of features, that you create once and then save in a library for future use. To help you get familiar with library features, SolidWorks has provided sample features for your use. You will first copy the desired features into the new folder that you created and then modify them for your specific needs.

In the **Design Library** tab, click on the plus sign next to the **design library** folder to expand the folder. Then, click on the plus sign next to the **features** folder to expand the folder.

Scroll the upper pane so that you can see your **Sheetmetal** folder.

Click on the **Sheetmetal** folder in the upper pane to view the contents of the folder in the lower pane.

In the lower pane, left click and drag **d-cutout**. Hold down the **Ctrl** key to copy the feature (the cursor should change to a plus sign). Move the cursor over the **Punching Shapes** folder and release the mouse button. If you accidentally move the file without copying, click on the folder you dragged it to. Then, re-drag the file while holding the **Ctrl** key to copy it back to the original folder.

In the lower pane, scroll down until you see **sw-a081**. Left click and drag **sw-a081** while holding the **Ctrl** key and drop it onto the **Punching Shapes** folder.

In the lower pane, scroll to the bottom until you see **tombstone relief**. Left click and drag **tombstone relief** while holding the **Ctrl** key and drop it onto the **Punching Shapes** folder.

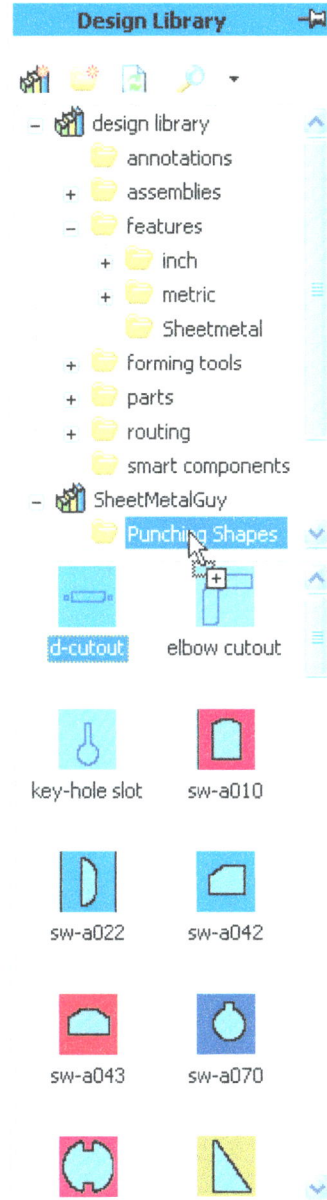

In the upper pane in the **features** folder, click on the plus sign next to the **inch** folder to expand the folder. Click on the **slots** folder to view the contents of the folder in the lower pane.

Scroll the upper pane so that you can see your **Punching Shapes** folder. Left click and drag **straight slot** while holding the **Ctrl** key and drop it onto the **Punching Shapes** folder.

Click on the **Punching Shapes** folder to view the contents of the folder in the lower pane.

Right click on **d-cutout** and pick **Rename** from the menu. Type 'RS-232' for the file name. The file names follow the standard for Windows. The complete path and file name and extension can contain up to 255 characters. File names cannot include a forward slash (/), backslash (\), greater than sign (>), less than sign (<), asterisk (*), question mark (?), quotation mark ("), pipe symbol (|), or colon (:). Rename **straight slot** to 'obround', and then rename **sw-a081** to '**DBL KWY .5**'. In SolidWorks 2007, use '**DBL KWY**' to avoid a file extension name problem.

In the **Design Library** tab, click the **Refresh** icon to refresh the view. The features are sorted alphabetically.

Edit an Existing Library Feature

The **End Condition** of some of the sample files is **Through All**. **Through All** extends the cut from the sketch plane through all existing geometry. In sheet metal, the typical cut should only go through the thickness of the material. In order to avoid cutting through all the flanges and features of the part, you must change the **End Condition** in your copied punching shapes.

To do this, double click on **RS-232** in the lower pane of the **Design Library** tab.

In the FeatureManager design tree, right click on **D Cut** and pick **Edit Feature** from the menu.

In the **D Cut** PropertyManager, change the **End Condition** to **Up To Next**, which extends the cut from the sketch plane to the next surface that intercepts the sketch profile.

Click the green check mark button at the top of the **D Cut** PropertyManager to accept the new settings.

Click on the lower **X** icon in the upper right corner of the SolidWorks program window to close the file, or pull down the "File" menu and pick **Close**.

When the **SolidWorks** dialog box appears, click the **Yes** button to save your changes to **RS-232**. Library features use the .sldlfp extension.

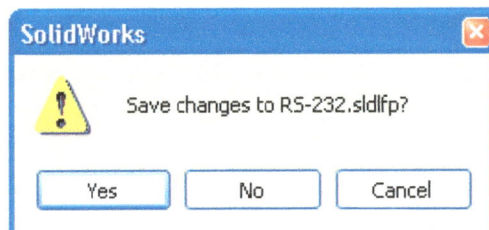

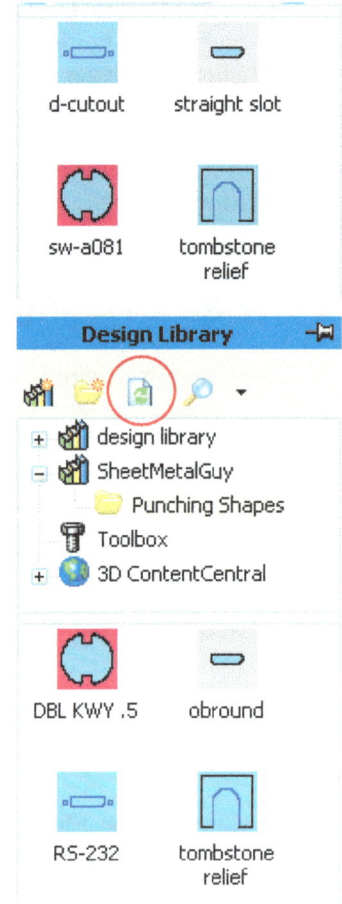

Click the **Design Library** icon in the Task Pane.

Click on the **Pin** icon in the title bar of the Task Pane to keep the Task Pane open while you are working. If the Task Pane is unpinned, it collapses when you drag an item into the graphics area, or when you open a new SolidWorks document. You can click on the **Pin** icon at any time to unpin the Task Pane.

Next, double click on **tombstone relief** in the lower pane of the **Design Library** tab.

In the FeatureManager design tree, right click on **Cut-Extrude1** and pick **Edit Feature** from the menu.

In the **Cut-Extrude1** PropertyManager, change the **End Condition** to **Up To Next**.

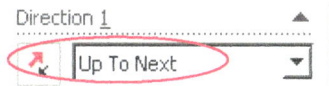

Click the green check mark button at the top of the **Cut-Extrude1** PropertyManager.

Click on the lower **X** icon in the upper right corner of the SolidWorks program window to close the file, or pull down the "File" menu and pick **Close**.

When the **SolidWorks** dialog box appears, click the **Yes** button to save your changes to **tombstone relief**.

Finally, double click on **obround** in the lower pane of the **Design Library** tab.

In the FeatureManager design tree, right click on **Slot** and pick **Edit Feature** from the menu.

In the **Slot** PropertyManager, change the **End Condition** to **Up To Next**.

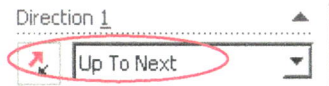

Click the green check mark button at the top of the **Slot** PropertyManager.

The **obround** uses a design table to give you a list of available sizes in one file. A design table allows you to build multiple configurations of parts or assemblies by specifying parameters in an embedded Microsoft Excel worksheet. To use design tables, you must have Microsoft Excel installed on your computer.

In the FeatureManager design tree, right click on **Design Table** and pick **Edit Table** from the menu.

The worksheet appears in the window. You can edit the table as needed, such as changing the parameter values in the cells, adding or deleting rows for additional configurations, or adding columns to control additional parameters.

Right click on the **3** row number and pick **Insert** from the menu to add a new row.

Double click in column A, row 3 to activate the cell. Type '**0.0625 X 0.75**' and then press the **Tab** key on the keyboard.

Type '**.0625**' and then press the **Tab** key on the keyboard.

Type '**.75**' and then press the **Enter** key on the keyboard.

You may add as many sizes as you wish to a design table. To delete any of the sizes, simply right click on the row number and pick **Delete** from the menu.

Click the green check mark at the top right of the graphics area to close the design table.

	A	B	C
		Width@SlotSketch	Length@SlotSketch
1			
2	0.094 X 1.5	0.094	1.5
3	0.0625 X 0.75	0.063	0.75
4	0.125 X 0.25	0.125	0.25
5	0.125 X 0.312	0.125	0.312

Click the **OK** button in the **SolidWorks** dialog box.

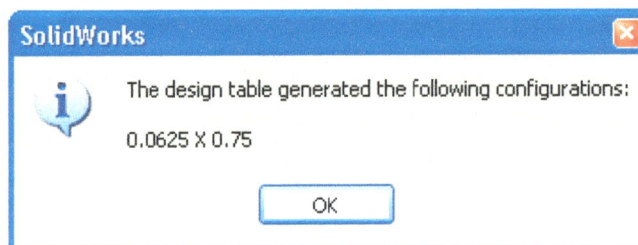

In the FeatureManager design tree, right click on **Slot** and pick **Edit Sketch** from the menu.

Press the **f** key to **Zoom to Fit**.

Delete the 3.000 and 2.000 dimensions. Then, click on the parallel relation (18) and delete it as well. This will make it easier for you to locate and orientate the punching shape when you place it into a part.

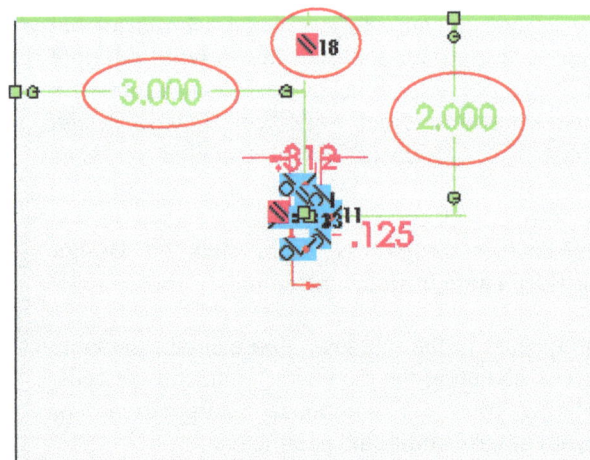

Exit the sketch by clicking the **Exit Sketch** icon in the CommandManager or in the upper right corner of the graphics area.

When you save a library feature or a forming tool, the thumbnail graphic reflects the view when the document is saved. Before saving, zoom in and orient the part or assembly so that the thumbnail graphic looks the way that you want it to look.

Click on the lower **X** icon in the upper right corner of the SolidWorks program window to close the file, or pull down the "File" menu and pick **Close**.

When the **SolidWorks** dialog box appears, click the **Yes** button to save your changes to **obround**.

Create a Custom Punching Shape

To create a new library feature, you first create a base feature to which you add the features that you want included in the library feature.

Begin a new **Part** document by clicking the **New** icon in the "Standard" toolbar, or pull down the "File" menu and pick **New**.

Create a base flange by clicking the **Sheet Metal** icon in the control area of the CommandManager. Then, click the **Base-Flange/Tab** icon from the toolbar, or pull down the "Insert" menu and pick **Sheet Metal – Base Flange**.

Select the **Front** plane when prompted to select a plane.

Create a rectangle with the lower left corner at the origin using the **Rectangle** icon in the CommandManager, or pull down the "Tools" menu and pick **Sketch Entities – Rectangle**.

Click the **Smart Dimension** icon in the CommandManager, or pull down the "Tools" menu and pick **Dimensions – Smart**.

Add a '**3**' vertical dimension to the left vertical line and a '**3**' horizontal dimension to the bottom horizontal line.

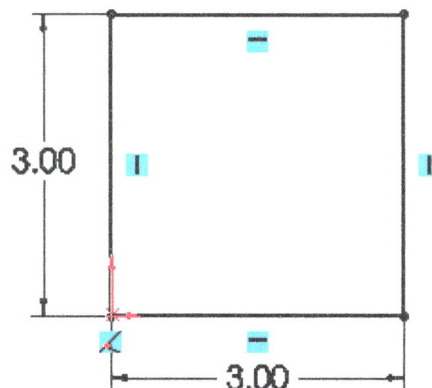

Exit the sketch by clicking the **Exit Sketch** icon in the CommandManager or in the upper right corner of the graphics area.

In the **Base Flange** PropertyManager under **Sheet Metal Parameters**, set the **Thickness** to '**.06**'. Make sure that the **Reverse direction** check box is not checked.

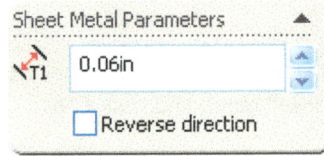

Click the green check mark button at the top of the **Base Flange** PropertyManager to accept the settings and create the part.

Click the **Extruded Cut** icon from the CommandManager or pull down the "Insert" menu and pick **Cut – Extrude**.

Select the front of the part as the plane onto which you will create the sketch.

In the bottom left corner of the graphics area, change the View orientation by clicking the pull down arrow and picking **Front**, or press **Ctrl-1**.

Create a circle in the middle of the part using the **Circle** icon in the CommandManager, or pull down the "Tools" menu and pick **Sketch Entities – Circle**.

Create a vertical line on the right side of the circle using the **Line** icon in the CommandManager, or pull down the "Tools" menu and pick **Sketch Entities – Line**. Make sure that the line starts and ends on the circle.

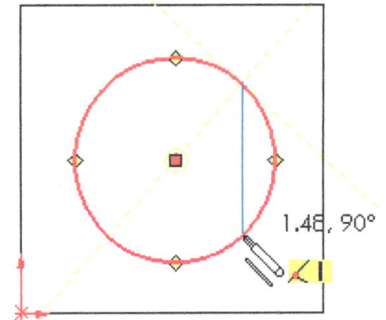

Next, click the **Trim Entities** icon or pull down the "Tools" menu and pick **Sketch Tools - Trim Entities**.

In the **Trim** PropertyManager, make sure that the **Power Trim** button is depressed.

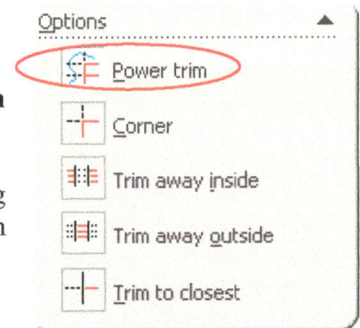

Click in the graphics area to the right of the circle and drag across the sketch entity to trim. A trail is created along the trim path as shown.

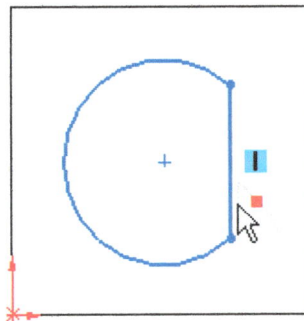

◇   Right click in the graphics area and pick **Smart Dimension** from the menu.

Add a '**.5**' radius dimension to the arc. Change the radius to a diameter dimension by right clicking on the R.5 dimension and picking **Properties** from the menu. In the **Dimension Properties** dialog box, check **Diameter dimension**. Click **OK**.

☑ Diameter dimension

☐ Display as linear dimension

◇   Next, add a dimension by clicking on the line. Then, press and hold the **Shift** key. Select the edge of the arc. Click and place a '**.875**' dimension, and then release the **Shift** key. If you release the **Shift** key in SolidWorks 2006 before placing the dimension, the dimension will be from the center of the circle. Rather than dimensioning to the center of the arc, the **Shift** key allowed you to dimension to the outside of the arc.

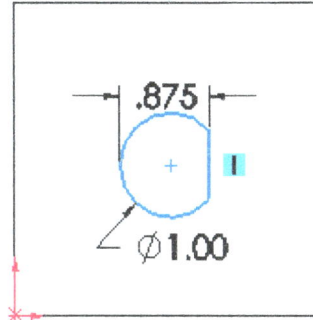

Press the **Escape** key to deselect **Smart Dimension**.

Click and drag the center point of the arc to move the sketch to the center of the part. You do not want to have any relations to the part since the shape will be used as a library feature.

✎   Exit the sketch by clicking the **Exit Sketch** icon in the CommandManager or in the upper right corner of the graphics area.

✔   In the **Cut-Extrude** PropertyManager, check the **Link to thickness** check box and then click the green check mark button at the top of the **Cut-Extrude** PropertyManager.

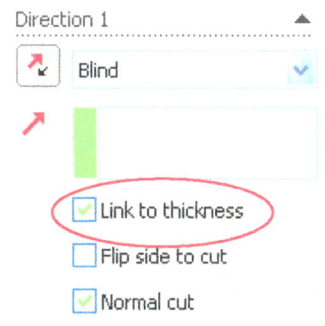

Pull down the "Tools" menu and pick **Options**.

On the **Document Properties** tab, select **Colors**.

In the **Model\Features** colors box, select **Library Feature**.

Click on the **Edit** button. In the **Color** dialog box, pick the yellow color (second column, second row) and click **OK**. In the **Document Properties** dialog box, click the **OK** button.

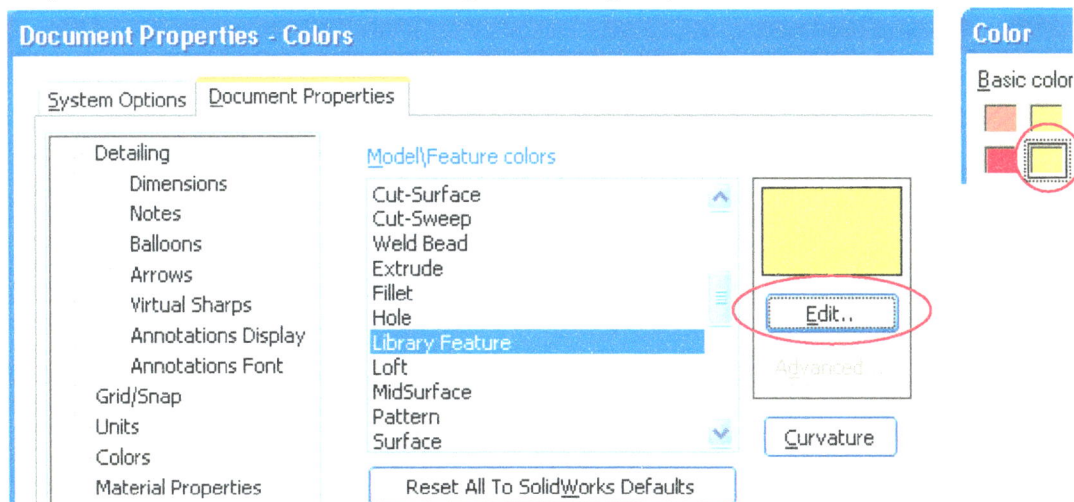

Each time you add the library feature to a part, the applied color is displayed.

Click and drag **Cut-Extrude1** to the lower pane of the **Design Library** tab. The cursor will change indicating that you are copying the feature. Release the mouse button. You may also select **Cut-Extrude1** in the FeatureManager design tree to highlight it. Then, pull down the "File" menu and pick **Save As**.

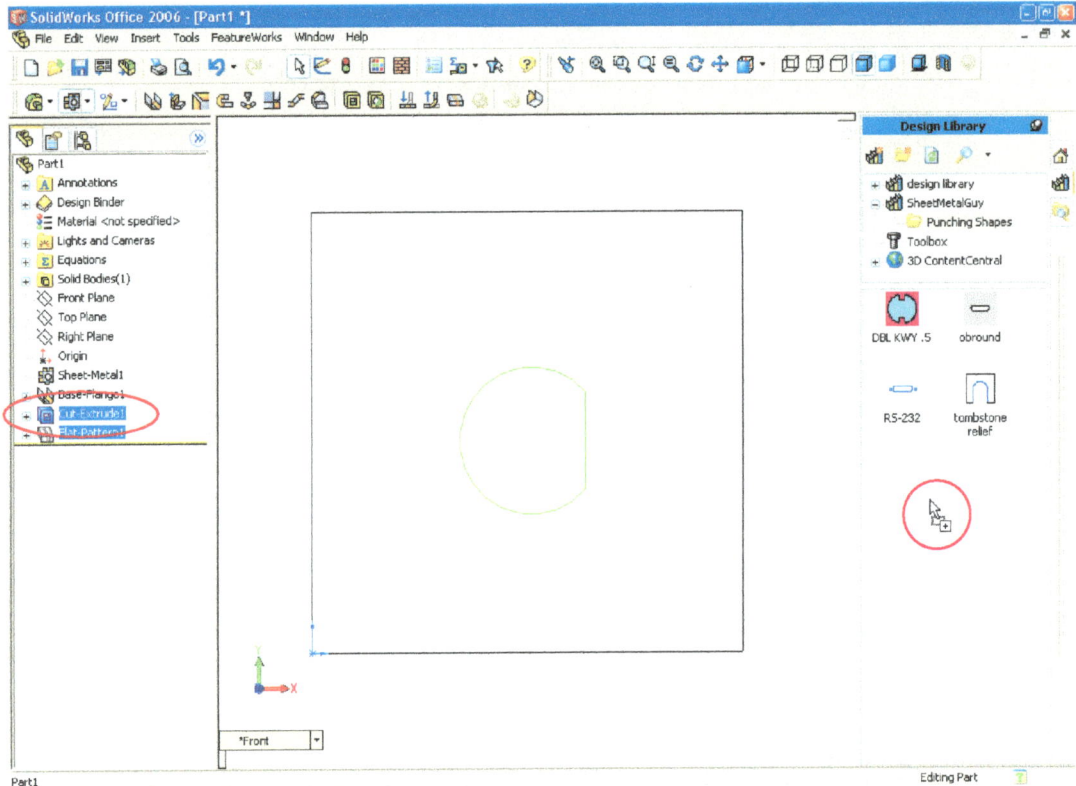

In the **Save As** dialog box, type '**SingleD**' for the **File name**.

Make sure **Save as type** is set to **Lib Feat Part (*.sldlfp)**. Library features have the .sldlfp extension.

Click **Save**, and then, in the **SolidWorks** dialog box, click **Yes**.

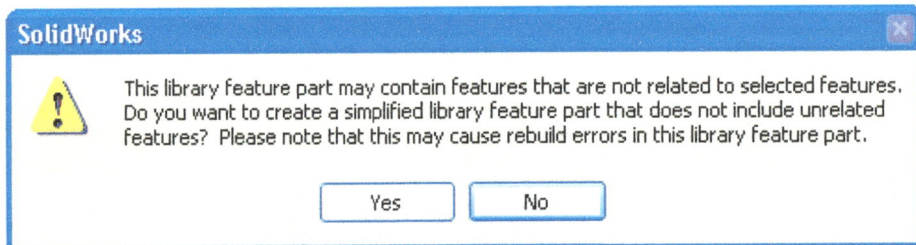

Click on the lower **X** icon in the upper right corner of the SolidWorks program window to close the file, or pull down the "File" menu and pick **Close**.

Copy an Existing Forming Tool

Forming tools act as dies that bend, stretch, or otherwise form sheet metal to create form features such as louvers, lances, flanges, and ribs. SolidWorks includes some sample forming tools to get you started.

In the **Design Library** tab, right click on **SheetMetalGuy** and pick **New Folder** from the menu. You may also click on **SheetMetalGuy** to highlight it and then click on the **Create New Folder** icon.

A new folder is created under **SheetMetalGuy**. Enter '**Forming Tools**' for the folder name.

The file name extension for forming tools is the same as a SolidWorks part document (.sldprt). Since SolidWorks handles forming tools differently from other parts, the software recognizes them by the folder classification.

Right click on the **Forming Tools** folder and pick **Forming Tools Folder** from the menu to tell SolidWorks that the folder contains forming tools.

In the **Design Library** tab, click on the plus sign next to the **design library** folder to expand the folder. Then, click on the plus sign next to the **forming tools** folder to expand the folder.

Click on the **embosses** folder in the upper pane to view the contents of the folder in the lower pane.

SolidWorks 2006 does not let you drag and drop forming tools to copy them to your new folder.

Instead, in the lower pane, double click on the **dimple** forming tool to open the file.

Pull down the "File" menu and pick **Save As**.

Browse to your new **Forming Tools** folder in the <install_dir>/**data/SheetMetalGuy** folder.

Click the **Save** button.

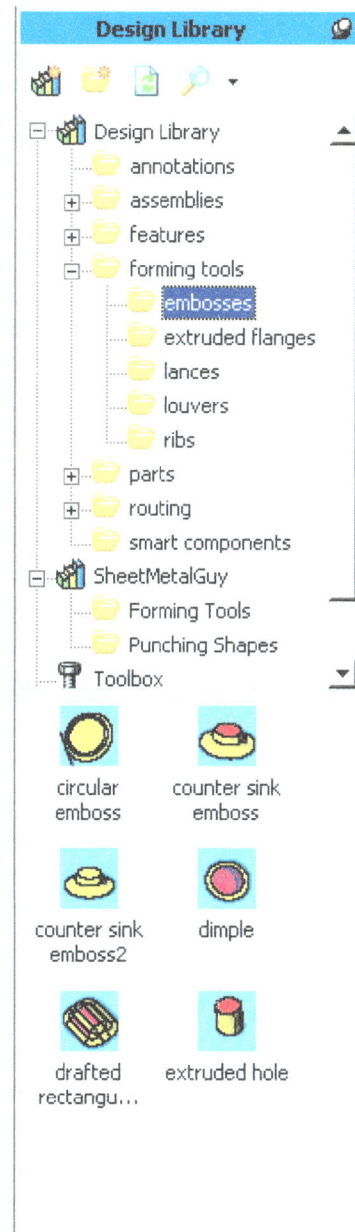

Pull down the "Tools" menu and pick **Options**.

On the **Document Properties** tab, select **Units**.

In the **Unit system** box, select **IPS (inch, pound, second)**.

Click the **OK** button.

In the FeatureManager design tree, right click on the **Annotations** folder and pick **Show Feature Dimensions**.

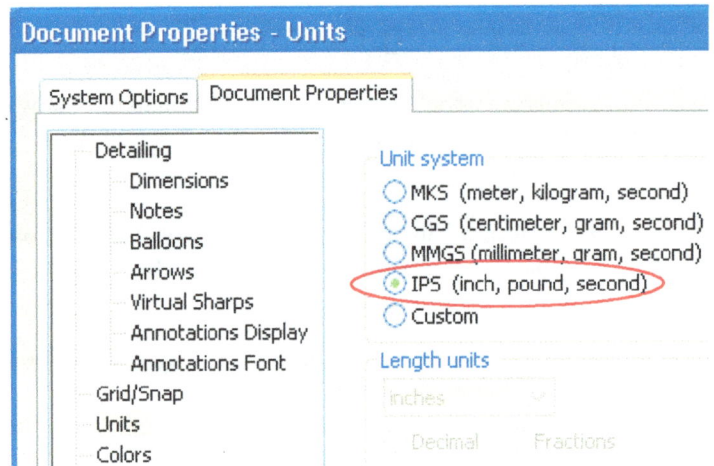

Document Properties - Units

System Options | Document Properties

Detailing
    Dimensions
    Notes
    Balloons
    Arrows
    Virtual Sharps
    Annotations Display
    Annotations Font
Grid/Snap
Units
Colors

Unit system
- MKS (meter, kilogram, second)
- CGS (centimeter, gram, second)
- MMGS (millimeter, gram, second)
- IPS (inch, pound, second)
- Custom

Length units
Inches

Decimal   Fractions

Since the part was created in metric, the dimensions are converted into inches.

Double click the dimensions and change them to the values shown. You may need to rotate the part to see all the dimensions.

Click the **Rebuild** icon in the "Standard" toolbar, or press **Ctrl-B**.

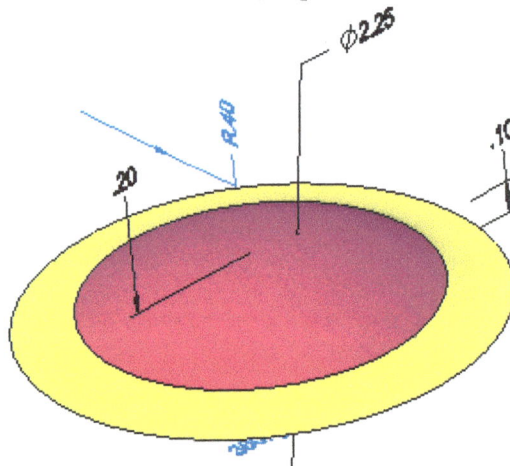

Once the dimensions have been updated, in the FeatureManager design tree, right click on the **Annotations** folder and pick **Show Feature Dimensions** again to hide the dimensions.

Click the **Save** icon in the "Standard" toolbar, or pick **Save** from the "File" pull down menu.

Click on the lower **X** icon in the upper right corner of the SolidWorks program window to close the file, or pull down the "File" menu and pick **Close**.

       SolidWorks for the Sheet Metal Guy

In the **forming tools** folder, under the **Design Library** folder, click on the **lances** folder in the upper pane to view the contents of the folder in the lower pane.

In the lower pane, double click on the **90 degree lance** forming tool.

Pull down the "File" menu and pick **Save As**.

Explore to your new **Forming Tools** folder in the <install_dir>/**data/SheetMetalGuy** folder.

In the **File name** box, type '**1 x .5 90 deg lance**'.

Click the **Save** button.

Pull down the "Tools" menu and pick **Options**.

On the **Document Properties** tab, select **Units**.

In the **Unit system** box, select **IPS (inch, pound, second)**.

Click the **OK** button.

Unit system
○ MKS (meter, kilogram, second)
○ CGS (centimeter, gram, second)
○ MMGS (millimeter, gram, second)
● IPS (inch, pound, second)
○ Custom

In the FeatureManager design tree, right click on **Lance-Tool** and pick **Edit Feature**.

Change the **Depth** to '**1**', and then, click the green check mark button at the top of the **Lance Tool** PropertyManager to accept the new settings.

In the FeatureManager design tree, right click on the **Annotations** folder and pick **Show Feature Dimensions**.

Change the height to '**.5**' and the radius to '**.125**'.

Click the **Rebuild** icon in the "Standard" toolbar, or press **Ctrl-B**.

Once the dimensions have been updated, in the FeatureManager design tree, right click on the **Annotations** folder and pick **Show Feature Dimensions** again to hide the dimensions.

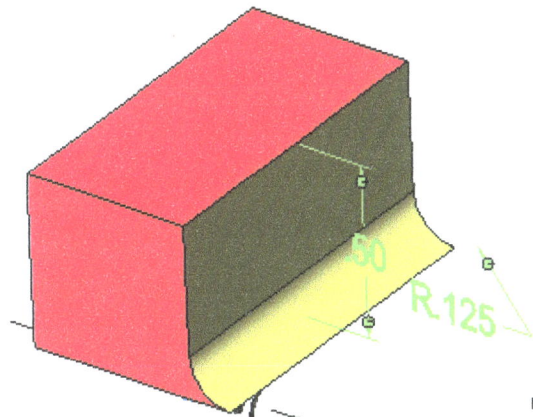

Click the **Save** icon in the "Standard" toolbar, or pick **Save** from the "File" pull down menu.

Click on the lower **X** icon in the upper right corner of the SolidWorks program window to close the file, or pull down the "File" menu and pick **Close**.

In the **forming tools** folder, click on the **louvers** folder in the upper pane to view the contents of the folder in the lower pane.

In the lower pane, double click on the **louver** forming tool.

Pull down the "File" menu and pick **Save As**.

Browse to your new **Forming Tools** folder in the <install_dir>/**data/SheetMetalGuy** folder. You may add as many folders as you want when organizing your forming tools. New folders may be added for multiple tools of the same type.

In the **File name** box, type '**2 x .375 louver**' and click the **Save** button.

Pull down the "Tools" menu and pick **Options**.

On the **Document Properties** tab, select **Units**.

In the **Unit system** box, select **IPS (inch, pound, second)**, and then click the **OK** button.

In the FeatureManager design tree, right click on **Base-Extrude** and pick **Edit Feature**.

Change the **Depth** to '**2.5**', and then, click the green check mark button at the top of the **Base Flange** PropertyManager to accept the new settings.

In the FeatureManager design tree, right click on the **Annotations** folder and pick **Show Feature Dimensions**.

Change the dimensions as shown, and then click the **Rebuild** icon in the "Standard" toolbar, or press **Ctrl-B**.

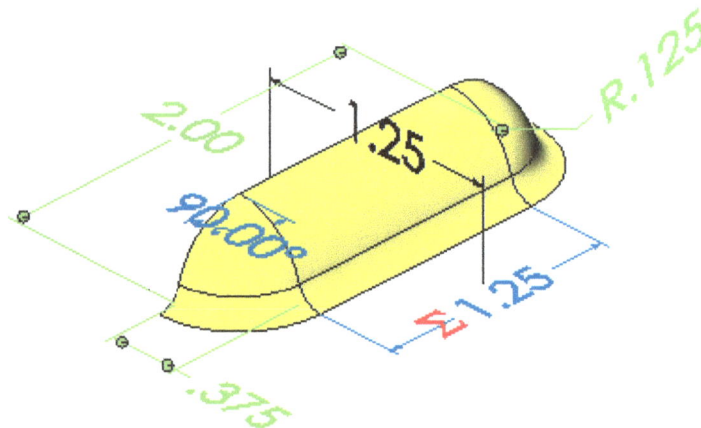

Once the dimensions have been updated, in the FeatureManager design tree, right click on the **Annotations** folder and pick **Show Feature Dimensions** again to hide the dimensions.

Click the **Save** icon in the "Standard" toolbar, or pick **Save** from the "File" pull down menu.

Click on the lower **X** icon in the upper right corner of the SolidWorks program window to close the file, or pull down the "File" menu and pick **Close**.

Click on the minus sign next to the **design library** folder to collapse the folder.

Click on your **Forming Tools** folder to display the contents of the folder in the lower pane. You may need to click on the **Refresh** icon.

Click on your **Punching Shapes** folder to display the contents of the folder in the lower pane.

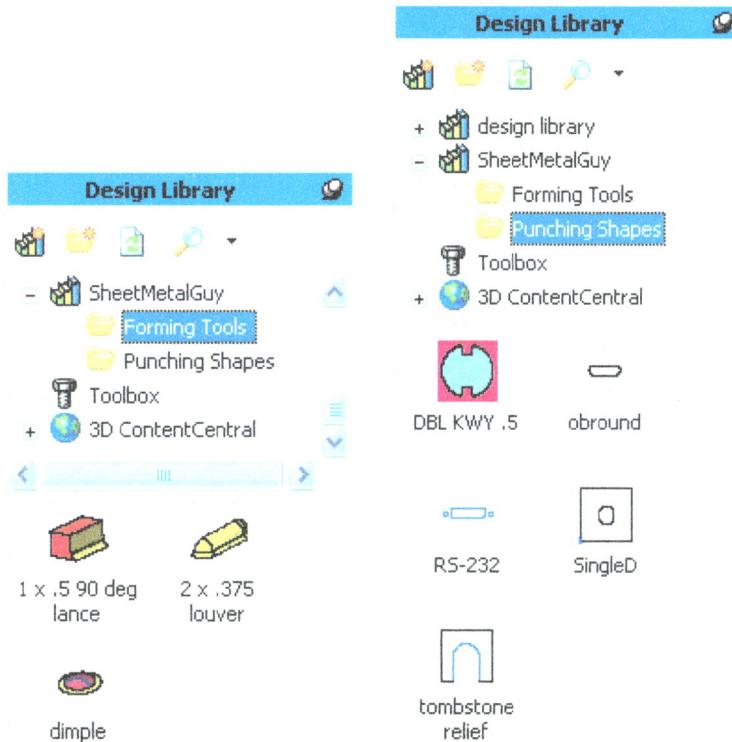

Create a New Forming Tool

You can create your own forming tools using many of the same steps you use to create any SolidWorks part. You can create forming tools and add them to sheet metal parts.

To create a forming tool, begin a new **Part** document by clicking the **New** icon in the "Standard" toolbar, or pull down the "File" menu and pick **New**.

Click the **Features** icon in the control area of the CommandManager. Then, click the **Extruded Boss/Base** icon in the toolbar, or pull down the "Insert" menu and pick **Boss/Base – Extrude**.

Select the **Top** plane when prompted to select a plane.

Create a rectangle with the origin inside the rectangle using the **Rectangle** icon in the CommandManager, or pull down the "Tools" menu and pick **Sketch Entities – Rectangle**.

Create a construction line diagonally across the rectangle by clicking the **Centerline** icon in the CommandManager, or pull down the "Tools" menu and pick **Sketch Entities – Centerline**.

Select the top left corner and then the bottom right corner to create the centerline.

Click the **Centerline** icon again to deselect the **Centerline** tool.

With the centerline still selected, hold down the **Ctrl** key and select the origin.

In the **Properties** PropertyManager, under **Add Relations**, click the **Midpoint** button.

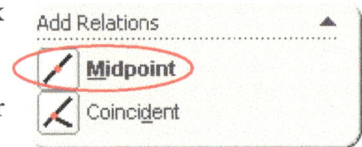

Click the **Smart Dimension** icon in the CommandManager, or pull down the "Tools" menu and pick **Dimensions – Smart**.

Add a '**1.5**' vertical dimension to the left vertical line and a '**3**' horizontal dimension to the bottom horizontal line.

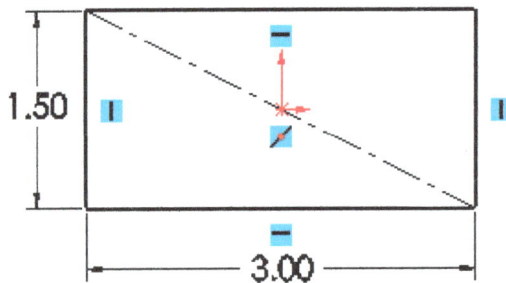

Exit the sketch by clicking the **Exit Sketch** icon in the CommandManager or in the upper right corner of the graphics area.

In the **Extrude** PropertyManager, click the **Reverse Direction** button. Under **Direction 1**, set the **Depth** to '**.125**'.

Click the green check mark button at the top of the **Extrude** PropertyManager to accept the settings and create the part.

Create a parallel plane by pulling down the "Insert" menu, and picking **Reference Geometry – Plane**.

Select **Front Plane** in the flyout FeatureManager design tree.

In the **Plane** PropertyManager, enter a **Distance** of '**.125**'.

Select the green check mark button at the top of the **Plane** PropertyManager.

With **Plane1** selected, click the **Extruded Boss/Base** icon from the toolbar, or pull down the "Insert" menu and pick **Boss/Base – Extrude**.

In the bottom left corner of the graphics area, change the View orientation by clicking the pull down arrow and picking **Normal To**.

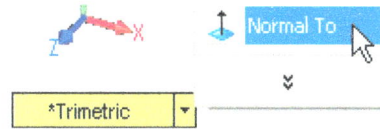

Create the sketch below using the **Line** icon in the CommandManager, or pull down the "Tools" menu and pick **Sketch Entities – Line**. You need to create four lines, including a bottom horizontal line, to form a closed loop shape.

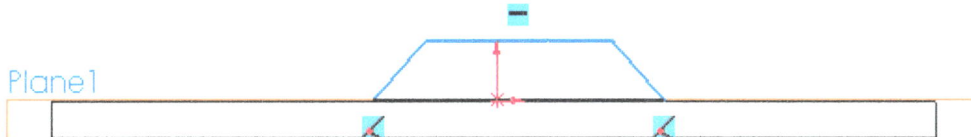

Press the **Escape** key to deselect the **Line** tool.

Select the bottom horizontal line of the sketch (the new line you just created, not the lower line of the original base feature), and then hold down the **Ctrl** key and select the origin point.

In the **Properties** PropertyManager, click the **Midpoint** button.

Select the left angled line, and then hold down the **Ctrl** key and select the right angled line.

In the **Properties** PropertyManager, click the **Equal** button.

Add the dimensions as shown in the sketch below using the **Smart Dimension** icon in the CommandManager, or pull down the "Tools" menu and pick **Dimensions – Smart**. The **0.25** dimension is not needed in SolidWorks 2007.

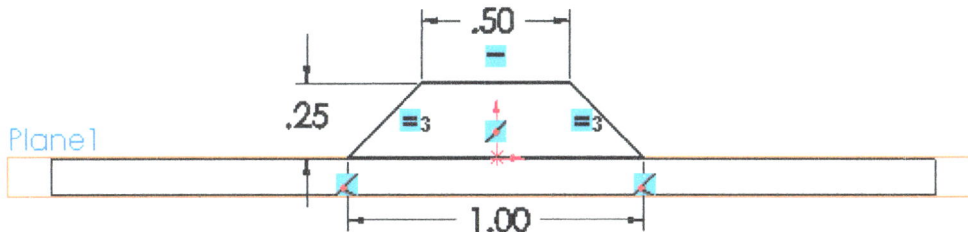

Exit the sketch by clicking the **Exit Sketch** icon in the CommandManager or in the upper right corner of the graphics area.

In the **Extrude** PropertyManager, under **Direction 1**, set the **End Condition** to **Mid Plane** and the **Depth** to '**.125**'.

Select the green check mark button at the top of the **Extrude** PropertyManager.

In the FeatureManager design tree, right click on **Plane1** and pick **Hide** from the menu.

In the bottom left corner of the graphics area, change the View orientation by clicking the pull down arrow and picking **Trimetric**.

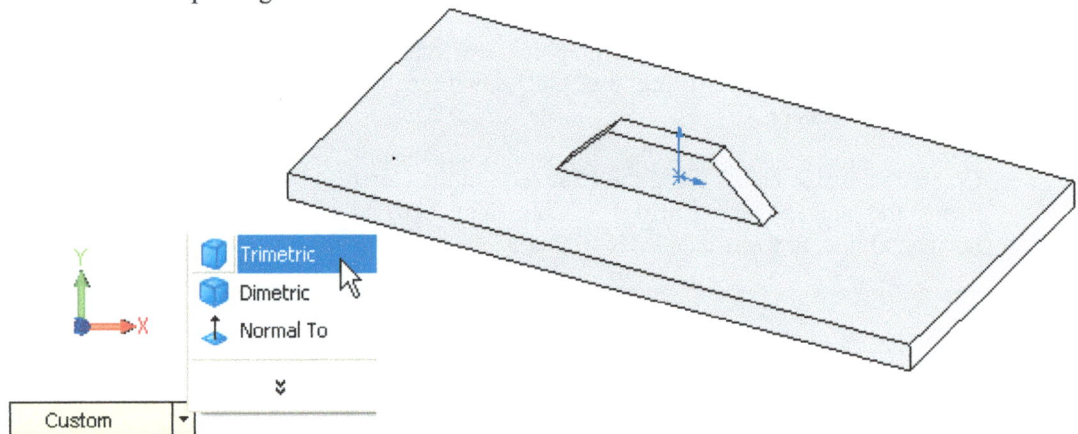

Click the **Features** icon in the control area of the CommandManager. Then, click the **Fillet** icon in the toolbar, or pull down the "Insert" menu and pick **Features – Fillet/Round**.

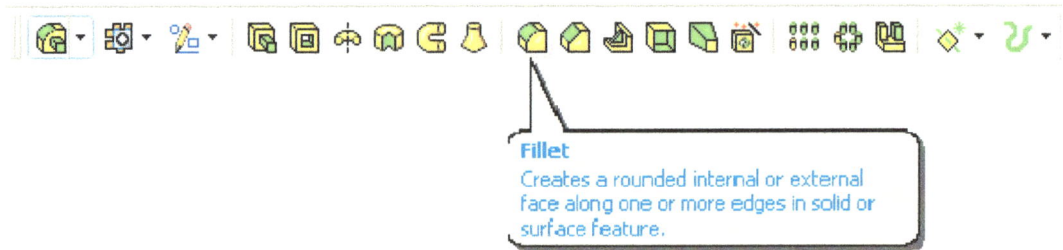

Fillet
Creates a rounded internal or external face along one or more edges in solid or surface feature.

In the **Fillet** PropertyManager, under **Items to Fillet**, enter '.125' for the **Radius**.

Select the four edge lines of **Extrude2** that you created in the previous step as shown. Rotate the part if necessary to select all four lines.

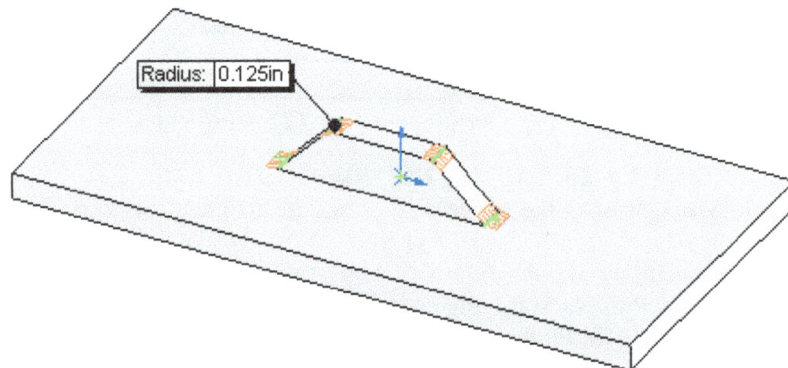

Radius: 0.125in

Select the green check mark button at the top of the **Fillet** PropertyManager.

Click the **Mirror** icon in the CommandManager, or pull down the "Insert" menu and pick **Pattern/Mirror – Mirror**.

In the flyout FeatureManager design tree, select **Front Plane** for the **Mirror Face/Plane**.

In the **Mirror** PropertyManager, click in the **Features to Mirror** box.

In the flyout FeatureManager design tree, select **Extrude2** and **Fillet1**. You should see a preview of ther second lance feature.

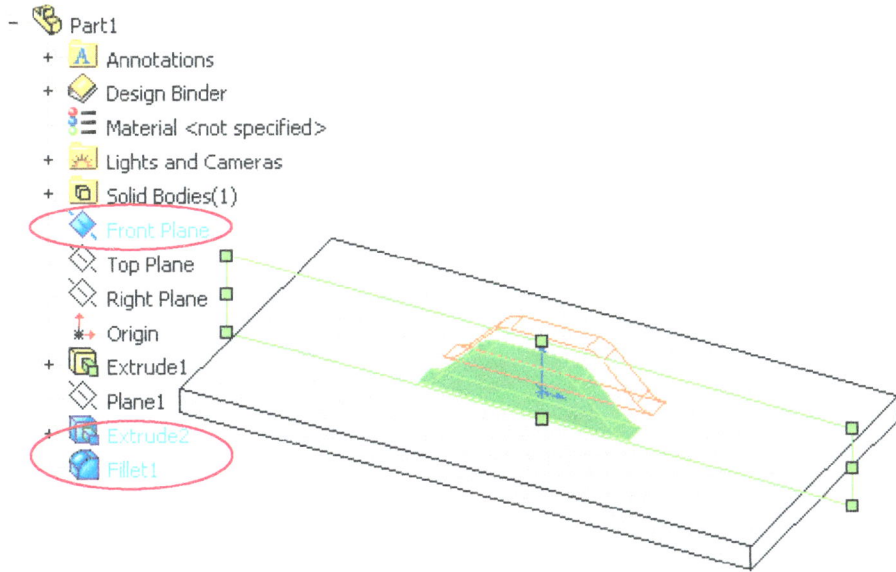

Click the green check mark button at the top of the **Mirror** PropertyManager.

In order to create a forming tool, pull down the "Insert" menu and pick **Sheet Metal – Forming Tool**.

In the graphics area, select the top of the part where the lances are located for the **Stopping Face**.

For the **Faces to Remove**, select the two sides of the part as shown.

Press the **left arrow** key five times to see the other side, and then select the other two faces of the other side as shown.

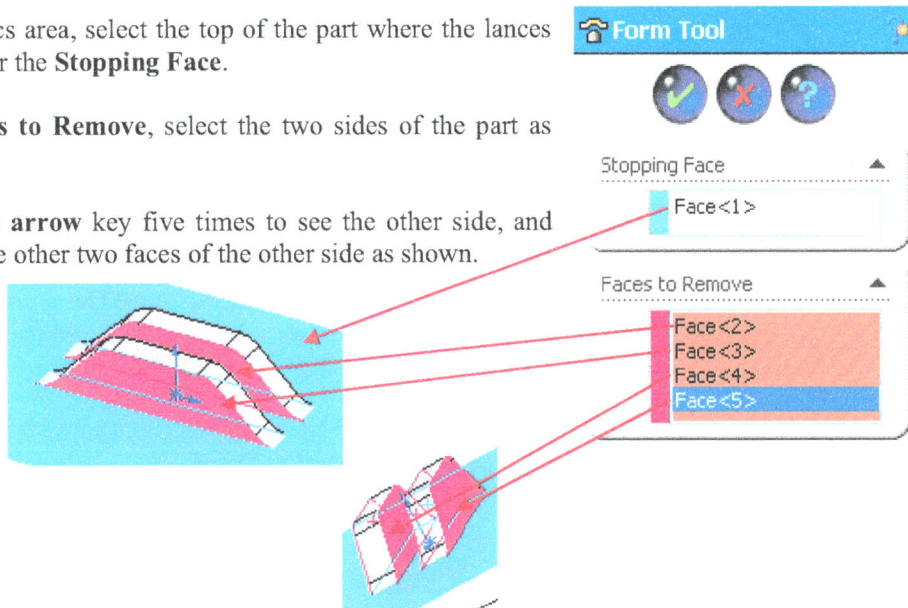

Click the green check mark button at the top of the **Form Tool** PropertyManager.

The Orientation Sketch and the colors that identify the stopping faces and the faces to remove are added automatically. Once you save the part to the forming tools folder in the Design Library, you can add the forming tool to a sheet metal part.

When you save a library feature or a forming tool, the thumbnail graphic reflects the view when the document was saved. Before saving, orient the part or assembly so that the thumbnail graphic looks the way you want it to look.

To do this, in the bottom left corner of the graphics area, change the View orientation by clicking the pull down arrow and picking **Trimetric**.

Then, **Zoom in** to show the double lance.

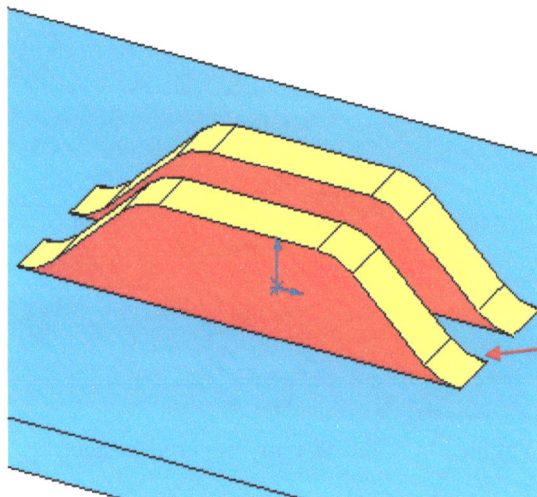

Saving the Part to the Forming Tool Folder

Click the **Save** icon in the "Standard" toolbar, or pick **Save** from the "File" pull down menu.

In the **Save As** dialog box, in the **File name** box, type '**.125 wide double lance**'.

Explore to your new **Forming Tools** folder in the <install_dir>/**data/SheetMetalGuy** folder and click the **Save** button.

In the Design Library, click on the plus to the left of the **SheetMetalGuy** folder if the folder is not expanded. Then, click on the **Forming Tools** folder. Click the **Refresh** icon to make certain everything is up to date. You should now have four forming tools in this folder.

Click on the **Punching Shapes** folder. You should have five shapes saved in this folder. These objects will be used in the upcoming chapters of this book.

Chapter 6

Cover

The Cover starts to use the shapes that you just placed in the Design Library. In this chapter, you will learn how to place and orientate these shapes.

The obround is retrieved from the Design Library and then placed into a line pattern.

Sometimes a special shape is needed but you do not feel it needs to be added to the Design Library. So, you can just create it on the fly, as is the case of the Double Keyway included here.

A Circular Fill Pattern can be a lot of work. SolidWorks offers a command that makes it very easy, as you will see.

Create the Base Flange

□ Begin a new **Part** document by clicking the **New** icon in the "Standard" toolbar, or pull down the "File" menu and pick **New**.

Create a base flange by clicking the **Sheet Metal** icon in the control area of the CommandManager. Then, click the **Base-Flange/Tab** icon from the toolbar, or pull down the "Insert" menu and pick **Sheet Metal – Base Flange**.

Select the **Right** plane when prompted to select a plane on which to sketch the feature cross-section.

Starting at the origin, create the sketch, shown below using the **Line** icon in the CommandManager, or pull down the "Tools" menu and pick **Sketch Entities – Line**.

Dimension the sketch as shown using the **Smart Dimension** icon in the CommandManager, or pull down the "Tools" menu and pick **Dimensions – Smart**.

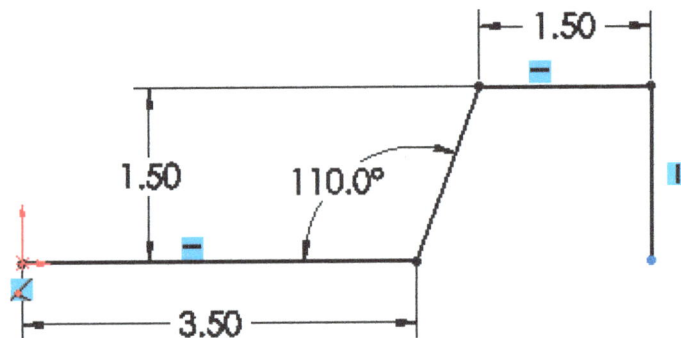

Exit the sketch by clicking the **Exit Sketch** icon in the CommandManager or in the upper right corner of the graphics area.

In the **Base Flange** PropertyManager under **Direction 1**, set the **End Condition** to **Blind** and the **Depth** to '**10**'. The origin should be on the right side of the part. If it is not, click the **Reverse Direction** button.

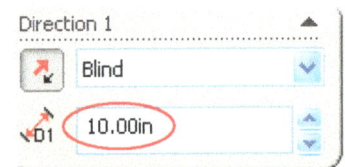

Under **Sheet Metal Parameters**, set the **Thickness** to '**.06**'. Make sure that the **Reverse direction** check box is checked.

Click the green check mark button at the top of the **Base Flange** PropertyManager to accept the settings and create the part.

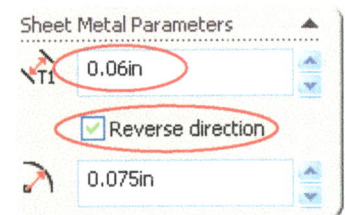

Change the display view to **Trimetric** by picking **Trimetric** in the View list in the lower left corner of the graphics area.

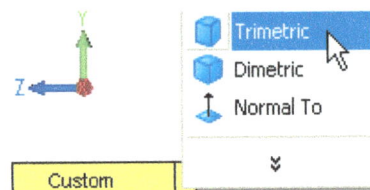

Create an Edge Flange

Click the **Sheet Metal** icon in the control area of the CommandManager. Then, click the **Edge Flange** icon in the CommandManager, or pull down the "Insert" menu and pick **Sheet Metal – Edge Flange**.

Select the right side of the part and move the cursor down and click to set the direction of the flange. Then, select the other side as shown.

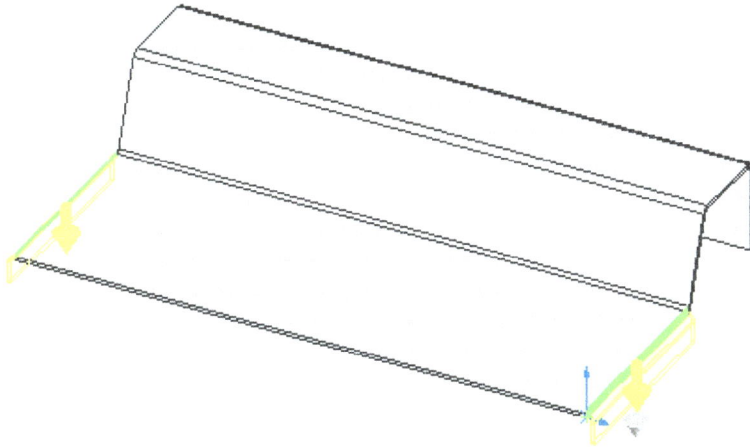

In the **Edge-Flange** PropertyManager, set the **Flange Length** to **Blind** and the **Length** to '.5'.

Click the **Outer Virtual Sharp** button and the **Bend Outside Flange Position** button.

Click the green check mark button at the top of the **Edge-Flange** PropertyManager to accept the settings and create the flange.

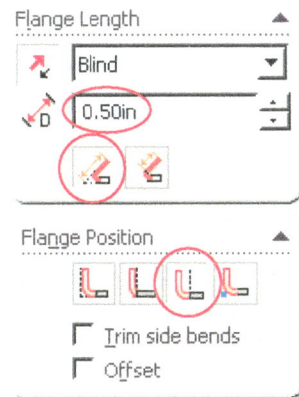

Flange Length

Blind

0.50in

Flange Position

Trim side bends

Offset

Add a Notch

Click the **Extruded Cut** icon from the CommandManager or pull down the "Insert" menu and pick **Cut – Extrude**.

Extruded Cut
Cuts a solid model by extruding a
sketched profile in one or two directions.

Select the top of the part for the plane to sketch on as shown.

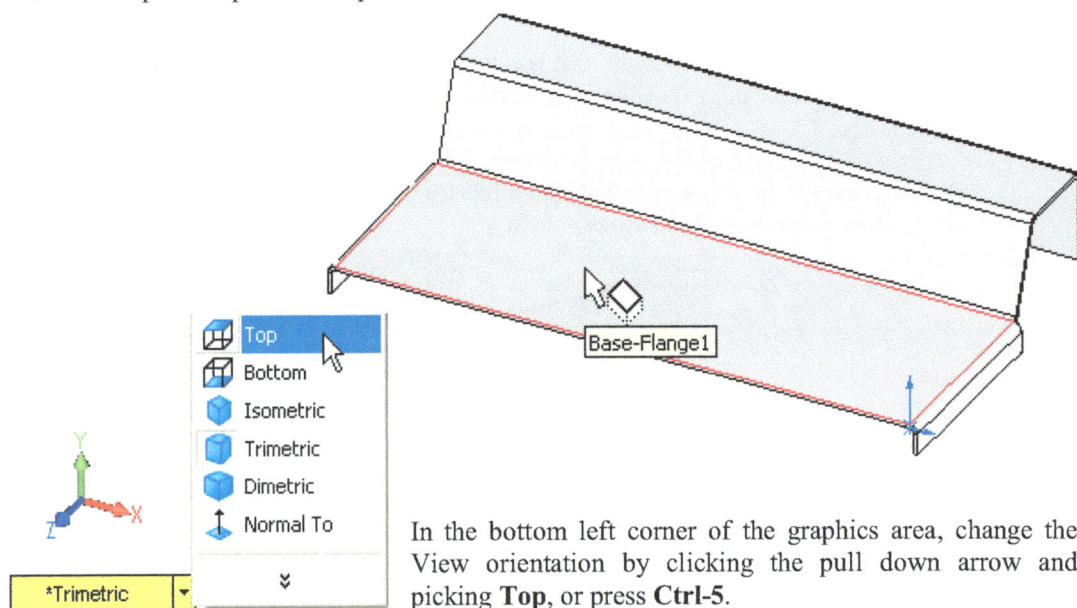

Top
Bottom
Isometric
Trimetric
Dimetric
Normal To

*Trimetric

Base-Flange1

In the bottom left corner of the graphics area, change the View orientation by clicking the pull down arrow and picking **Top**, or press **Ctrl-5**.

Create a rectangle as shown below using the **Rectangle** icon in the CommandManager, or pull down the "Tools" menu and pick **Sketch Entities – Rectangle**.

Dimension the sketch as shown using the **Smart Dimension** icon in the CommandManager, or pull down the "Tools" menu and pick **Dimensions – Smart**.

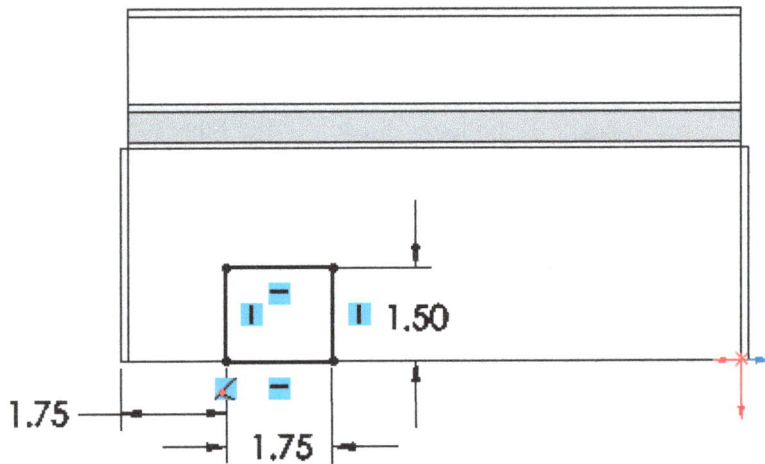

1.50

1.75

1.75

In the **Cut-Extrude** PropertyManager, check the **Link to thickness** check box.

Then, click the green check mark button at the top of the **Cut-Extrude** PropertyManager.

Direction 1

Blind

☑ Link to thickness
☐ Flip side to cut
☑ Normal cut

Add a Fill Pattern with a Fill Boundary

Click the **Sketch** icon in the control area of the CommandManager. Then, click the **Sketch** icon from the CommandManager, or pull down the "Insert" menu and pick **Sketch**.

Select the top of the part as shown.

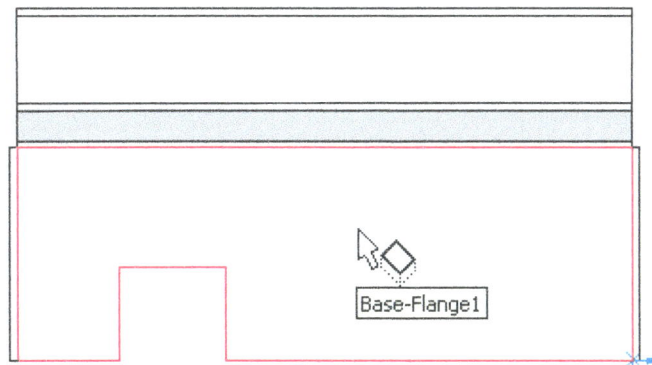

Create a centerline line from the bottom right corner of the notch to the upper right corner of the top of the face as shown using the **Centerline** icon in the CommandManager, or pull down the "Tools" menu and pick **Sketch Entities – Centerline**.

Create a circle at the midpoint of the centerline as shown below using the **Circle** icon in the CommandManager, or pull down the "Tools" menu and pick **Sketch Entities – Circle**.

Add a '**2.5**' dimension to the circle using the **Smart Dimension** icon in the CommandManager, or pull down the "Tools" menu and pick **Dimensions – Smart**.

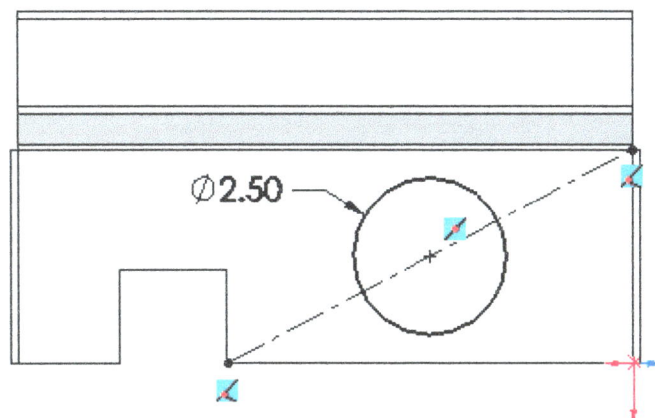

Exit the sketch by clicking the **Exit Sketch** icon in the CommandManager or in the upper right corner of the graphics area.

Pull down the "Insert" menu and pick **Pattern/Mirror – Fill Pattern**.

In the upper left hand corner of the graphics area, click on the plus sign next to the document name (Part1) to expand the flyout FeatureManager design tree.

In the **Fill Pattern** PropertyManager, click in the **Fill Boundary** box.

In the graphics area, select **Sketch5**.

In the **Fill Pattern** PropertyManager, under **Pattern Layout**, click the **Circular** button.

Set the **Loop Spacing** to '**.375**' and the **Instance Spacing** to '**.4375**'.

Click in the **Pattern Direction** box. Then, in the graphics area, click on the bottom edge of the part.

In the **Fill Pattern** PropertyManager, under **Features to Pattern**, click the **Create seed cut** radio button. Then, set the **Diameter** to '**.25**'.

Click the green check mark button at the top of the **Fill Pattern** PropertyManager to accept the settings.

In the FeatureManager design tree, right click on **Sketch5** and pick **Hide**.

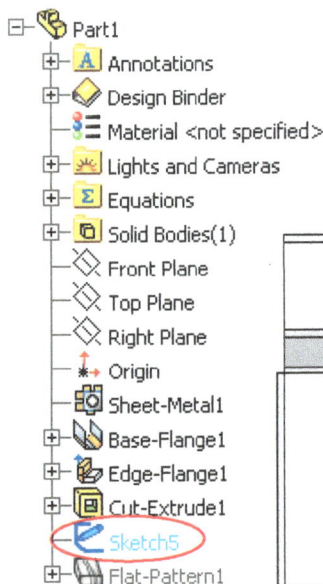

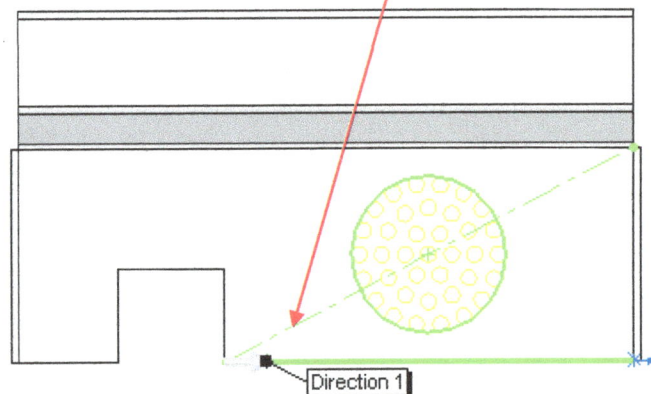

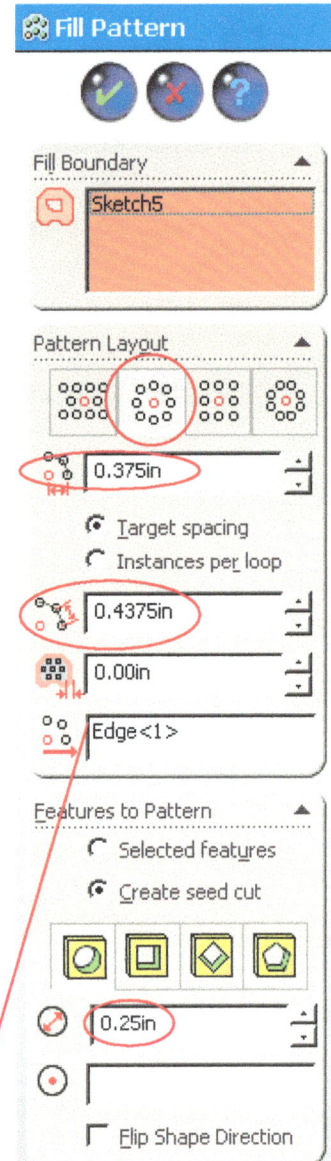

Add a Punching Shape from the Design Library

With the Task Pane shown, click the **Design Library** icon in the Task Pane.

In the **SheetMetalGuy** Design Library, click on your Punching Shapes folder to display the folder contents in the lower pane of the **Design Library** tab.

Drag the **obround** from the lower pane of the **Design Library** tab onto the top left of the part as shown.

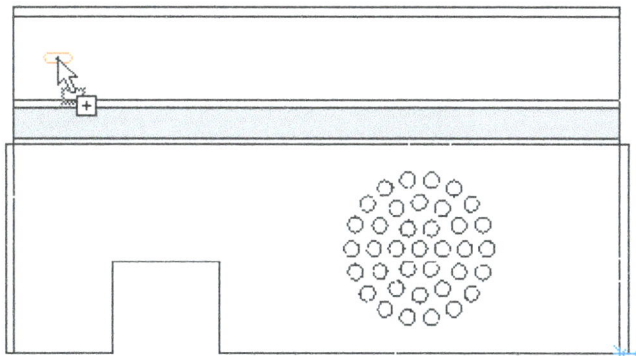

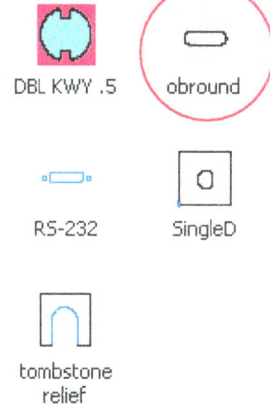

In the **obround** PropertyManager, scroll down the **Configuration** box to the bottom and pick '**0.0625 X 0.75**' from the list.

Under **Location**, click the **Edit Sketch** button.

To rotate the sketch, pull down the "Tools" menu and pick **Sketch Tools – Modify**.

In the **Modify Sketch** dialog box, under **Rotate**, enter '**90**' and press the **Enter** key on the keyboard. The obround sketch should rotate 90 degrees.

In the **Modify Sketch** dialog box, click the **Close** button.

Click the **Smart Dimension** icon in the CommandManager, or pull down the "Tools" menu and pick **Dimensions – Smart**.

Add a '.5' horizontal dimension from the left edge to the centerline of the obround. Then, add a '.75' vertical dimension from the top edge of the part to the midpoint of centerline as shown. You may need to zoom in to select the appropriate entities.

In the **Library Feature Profile** dialog box, click the **Finish** button.

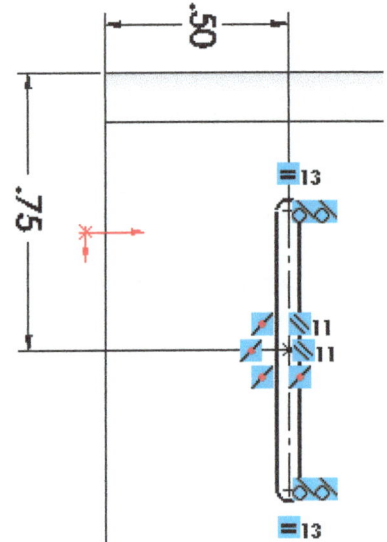

Create a Linear Pattern

Click the **Features** icon in the control area of the CommandManager. Then, click the **Linear Pattern** icon from the CommandManager, or pull down the "Insert" menu and pick **Pattern/Mirror – Linear Pattern**.

Linear Pattern
Patterns features, faces, and bodies in one or two linear directions.

Select the top horizontal line of the part near the left end to set the direction of the pattern.

In the **Linear Pattern** PropertyManager, set the **Spacing** to '**.375**'.

Set the **Number of Instances** to '**25**'.

Click in the **Features to Pattern** box.

In the upper left hand corner of the graphics area, click on the plus sign next to the document name (Part1) to expand the flyout FeatureManager design tree.

In the flyout FeatureManager design tree, select **obround <1>(0.0625 X 0.75)**.

Click the green check mark button at the top of the **Linear Pattern** PropertyManager.

Linear Pattern

Direction 1
Edge<1>
0.375in
25

Direction 2

Features to Pattern
obround<1>

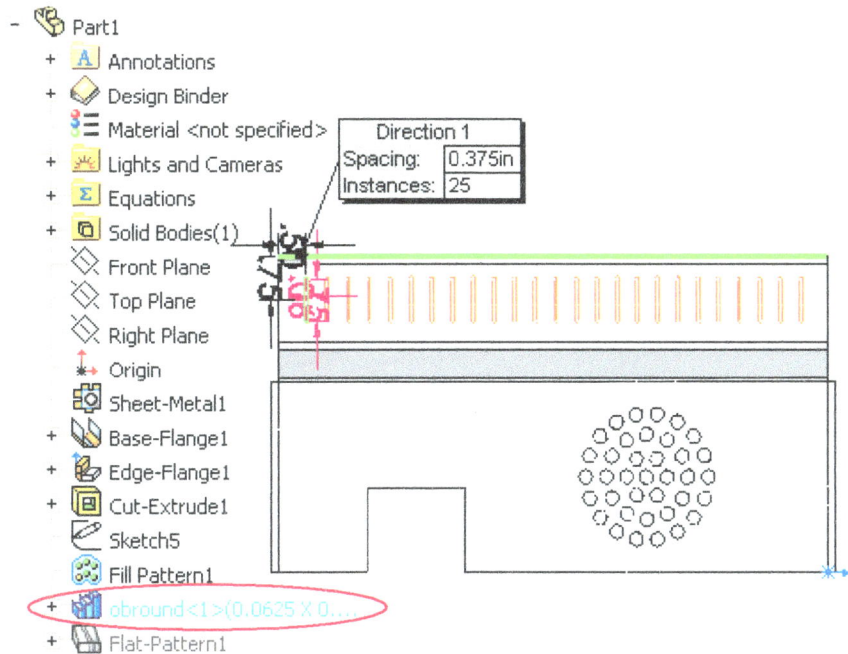

In the bottom left corner of your graphics area, click the pull down arrow and pick **Trimetric**.

*Trimetric

Add a RS-232 from the Design Library

Drag **RS-232** from the lower pane of the **Design Library** tab onto the front of the part as shown.

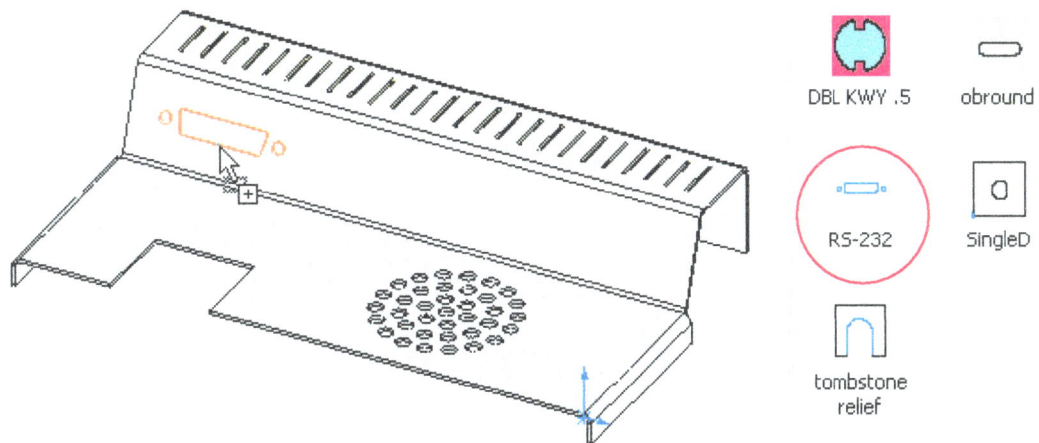

In the **RS-232** PropertyManager, under **Location**, click the **Edit Sketch** button.

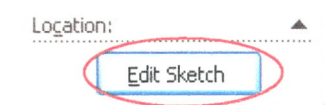

Location:

Edit Sketch

✏️ Click the **Smart Dimension** icon in the CommandManager, or pull down the "Tools" menu and pick **Dimensions – Smart**.

Add a '**2.25**' horizontal dimension from the left edge to the centerline of the RS-232. Then, add a '**.75**' vertical dimension from the bottom face edge line to the center of the left circle as shown. You may need to zoom in to select the appropriate entities.

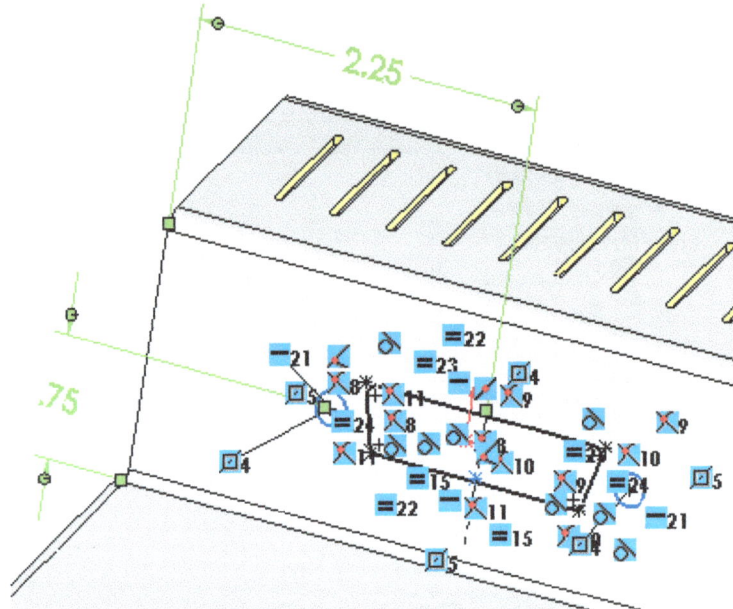

In the **Library Feature Profile** dialog box, click the **Finish** button.

Add a Double Keyway from the Design Library

Drag **DBL KWY .5** from the lower pane of the **Design Library** tab onto the front of the part.

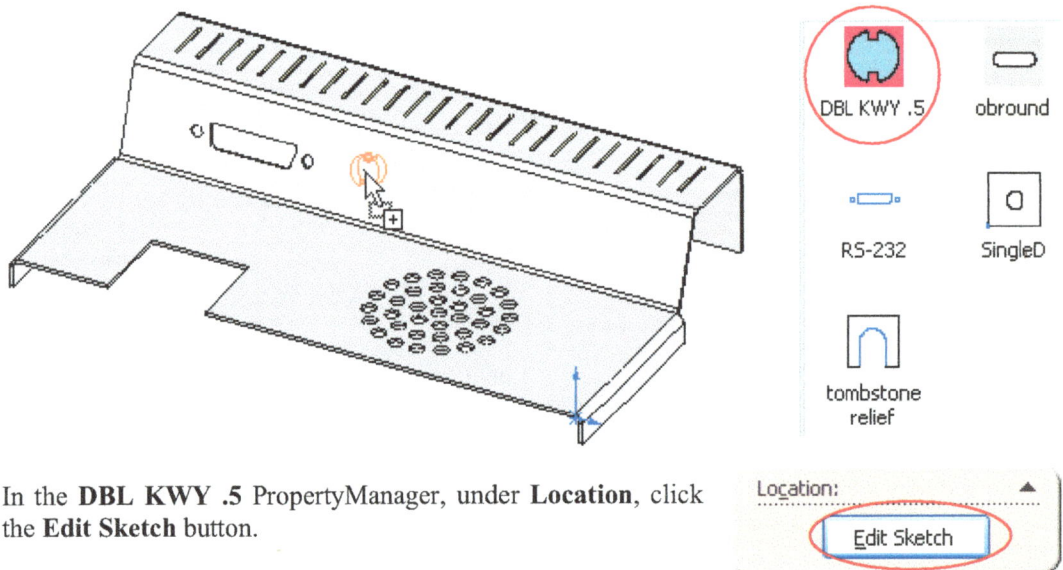

In the **DBL KWY .5** PropertyManager, under **Location**, click the **Edit Sketch** button.

Click the **Smart Dimension** icon in the CommandManager, or pull down the "Tools" menu and pick **Dimensions – Smart**.

Add a '**1.00**' horizontal dimension from the right circle of the RS-232 to the vertical centerline of the double keyway. Then, add a '**.75**' vertical dimension from the bottom face edge line to the horizontal centerline as shown. You may need to zoom in to select the appropriate entities.

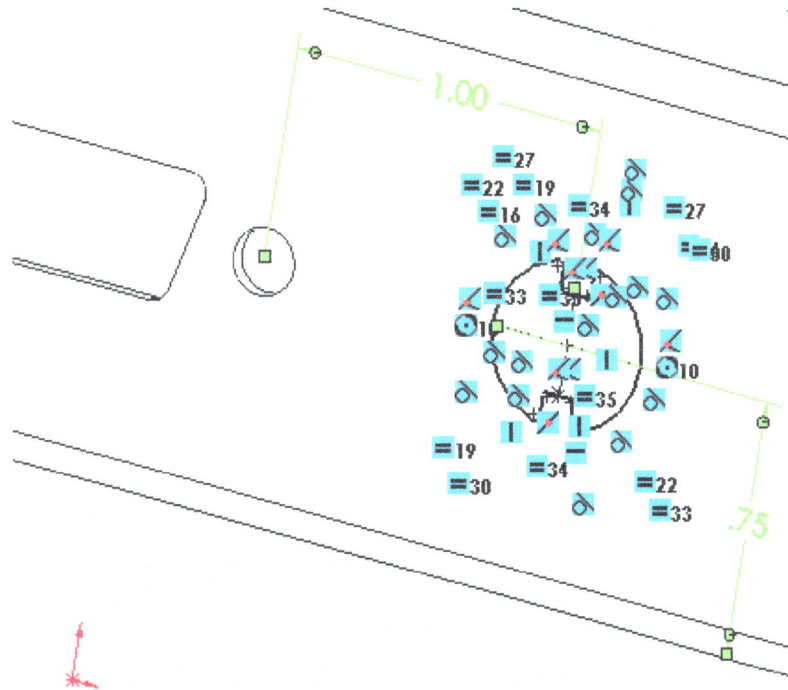

In the **Library Feature Profile** dialog box, click the **Finish** button.

Add a Custom Cutout

Click the **Sheet Metal** icon in the control area of the CommandManager. Then, click the **Extruded Cut** icon from the CommandManager or pull down the "Insert" menu and pick **Cut – Extrude**.

Extruded Cut
Cuts a solid model by extruding a sketched profile in one or two directions.

Press the **f** key to **Zoom to Fit**.

Select the front of the part for the plane to sketch on as shown.

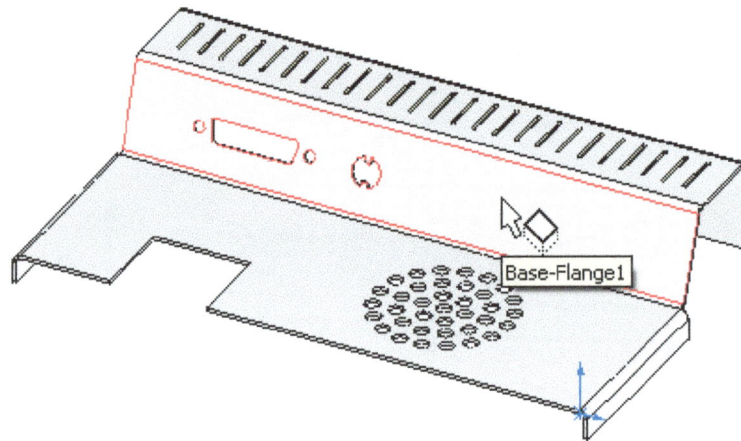

In the bottom left corner of the graphics area, change the View orientation by clicking the pull down arrow and picking **Normal To**.

Create a rectangle on the right side of the part as shown below using the **Rectangle** icon in the CommandManager, or pull down the "Tools" menu and pick **Sketch Entities – Rectangle**.

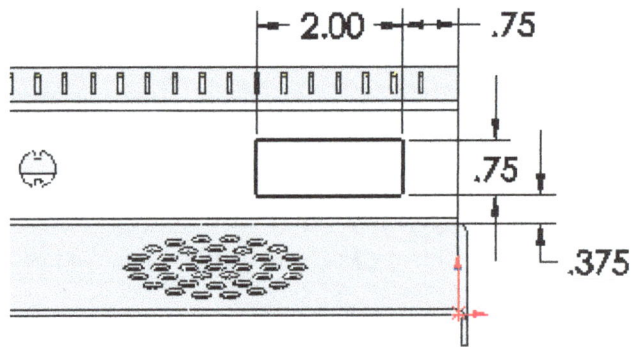

Dimension the sketch as shown using the **Smart Dimension** icon in the CommandManager, or pull down the "Tools" menu and pick **Dimensions – Smart**.

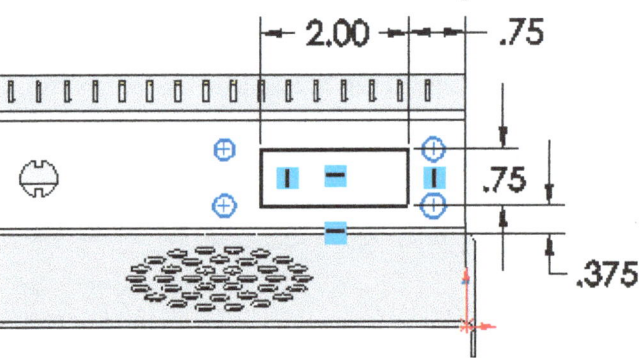

Create four circles in line with the top and bottom of the rectangle as shown below using the **Rectangle** icon in the CommandManager, or pull down the "Tools" menu and pick **Sketch Entities – Rectangle**.

Press the **Escape** key to deselect the **Circle** tool.

Select the four circles while holding the **Ctrl** key. Make sure that you select the circles and not he centerpoints.

In the **Properties** PropertyManager, click the **Equal** button.

Click the upper left hand corner of the rectangle. (The corner point should highlight). Hold down the **Ctrl** key and select the centerpoints of the top two circles.

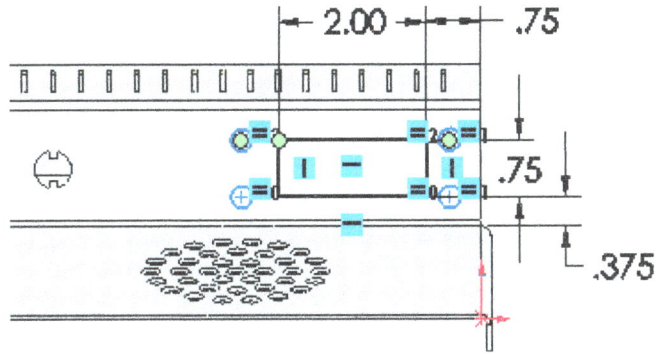

In the **Properties** PropertyManager, click the **Horizontal** button.

Click the lower left hand corner of the rectangle (the corner point should highlight). Hold down the **Ctrl** key and select the centerpoints of the bottom two circles. Then, in the **Properties** PropertyManager, click the **Horizontal** button.

Click the centerpoint of the top left circle. Hold down the **Ctrl** key and select the centerpoint of the lower left circle. Then, in the **Properties** PropertyManager, click the **Vertical** button.

Click the centerpoint of the top right circle. Hold down the **Ctrl** key and select the centerpoint of the lower right circle. Then, in the **Properties** PropertyManager, click the **Vertical** button.

Add a '**5/32**' diameter dimension to one of the circles, a '**3/16**' dimension between the right vertical line of the rectangle and the upper right circle, and a '**2.36**' dimension between the top left circle and the top right circle using the **Smart Dimension** icon in the CommandManager, or pull down the "Tools" menu and pick **Dimensions – Smart**.

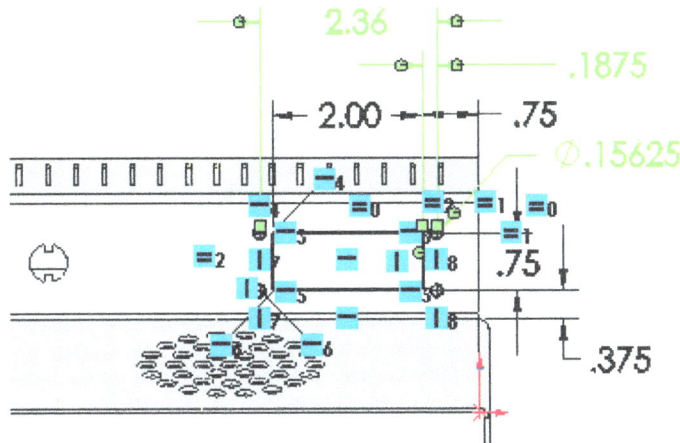

Exit the sketch by clicking the **Exit Sketch** icon in the CommandManager or in the upper right corner of the graphics area.

In the **Cut-Extrude** PropertyManager, check the **Link to thickness** check box.

Then, click the green check mark button at the top of the **Cut-Extrude** PropertyManager.

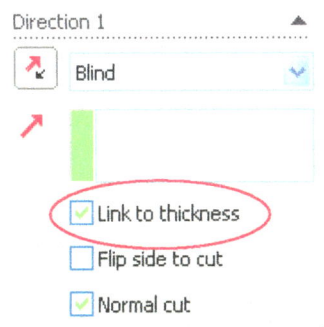

Edit the Notch and Add a Cutout

In the FeatureManager design tree, right click on **Cut-Extrude1**, and pick **Edit Sketch**.

Change the display view to **Top** by picking **Top** in the View list in the lower left corner of the graphics area.

Create a rectangle as shown above the notch rectangle using the **Rectangle** icon in the CommandManager, or pull down the "Tools" menu and pick **Sketch Entities – Rectangle**.

Dimension the sketch as shown using the **Smart Dimension** icon in the CommandManager, or pull down the "Tools" menu and pick **Dimensions – Smart**.

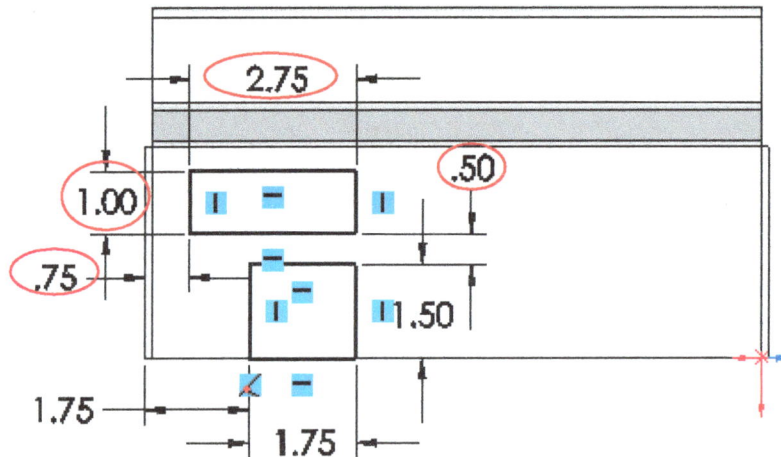

Exit the sketch by clicking the **Exit Sketch** icon in the CommandManager or in the upper right corner of the graphics area.

Change the display view to **Trimetric** by selecting **Trimetric** in the View list in the lower left corner of the graphics area.

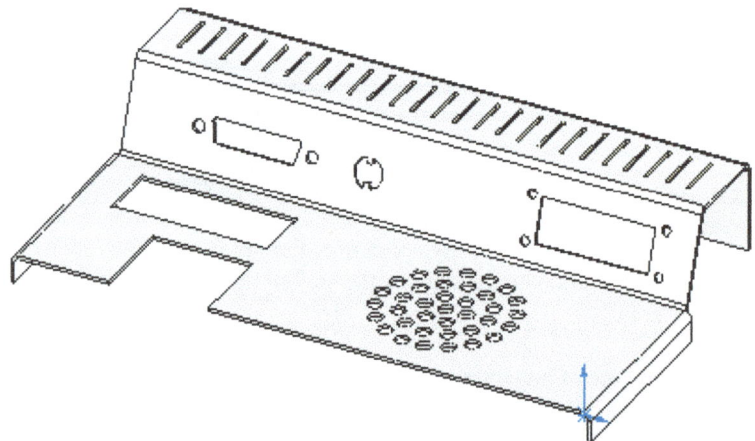

Saving the Part

Click the **Save** icon in the "Standard" toolbar, or pick **Save** from the "File" pull down menu.

In the **Save As** dialog box, in the **File name** box, type 'Cover'. Make sure that the **Save in** directory is your SolidWorks folder, not the Forming Tools folder. Click **Save**.

Chapter 7

Damper Channel

A family of parts is very common. Yet, to truly create a part file, which correctly updates for the different sizes of the part, may require more than just a few formulas. This simple channel will demonstrate how to use a Design Table and some basic "IF" logic to maintain the design intent.

The variables for this channel are the height, width, and length. The hole patterns must adjust accordingly. There are actually two different channels required to make the rectangular damper frame. This is one of the two channels.

The flange holes are basically on 4" centers, but the end holes must also line up with the holes of the second channel. Any odd spacing is placed between the first and second holes from each end. Oh, and do not forget that there is a minimum distance you will allow between these holes. That is where the "IF" logic comes into play.

You could put every dimension into the Design Table. This would allow you to totally control the part from the table. However, this being the first venture into a table, it might become overwhelming.

Create the Base Flange

Begin a new **Part** document by clicking the **New** icon in the "Standard" toolbar, or pull down the "File" menu and pick **New**.

Click on the **Pin** icon in the title bar of the Task Pane to collapse the Task Pane while you are working.

Create a base flange by clicking the **Sheet Metal** icon in the control area of the CommandManager. Then, click the **Base-Flange/Tab** icon from the toolbar, or pull down the "Insert" menu and pick **Sheet Metal – Base Flange**.

Select the **Right** plane when prompted to select a plane on which to sketch the feature cross-section.

Click on the **Line** icon in the CommandManager, or pull down the "Tools" menu and pick **Sketch Entities – Line**.

Create three lines forming the letter 'C' around the origin.

Click on the **Line** icon again to deselect it or you can press the **Escape** key on the keyboard.

Select the vertical line. Then, hold down the **Ctrl** key and select the origin.

In the **Properties** PropertyManager, click on the **Midpoint** button.

Select the top horizontal line. Then, hold down the **Ctrl** key and select the bottom line.

In the **Properties** PropertyManager, click on the **Equal** button.

Dimension the sketch as shown using the **Smart Dimension** icon in the CommandManager, or pull down the "Tools" menu and pick **Dimensions – Smart**.

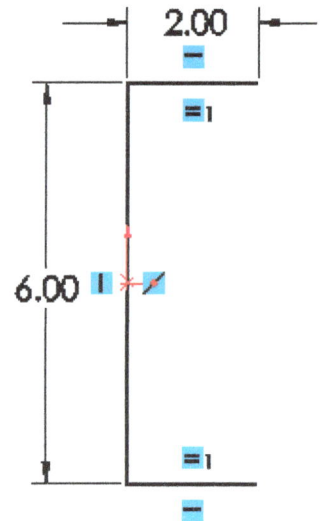

Exit the sketch by clicking the **Exit Sketch** icon in the CommandManager or in the upper right corner of the graphics area.

In the **Base Flange** PropertyManager under **Direction 1**, set the **End Condition** to **Blind** and the **Depth** to '30'.

Under **Sheet Metal Parameters**, set the **Thickness** to '.06'. Make sure that the **Reverse direction** check box is not checked. You want your dimensions to be to the outside of the part.

Click the green check mark button at the top of the **Base Flange** PropertyManager to accept the settings and create the part.

Create a Simple Hole

Click the **Sheet Metal** icon in the control area of the CommandManager. Then, click the **Simple Hole** icon in the toolbar, or pull down the "Insert" menu and pick **Features – Hole – Simple**.

Simple Hole
Creates a cylindrical hole on a planar face.

When prompted in the PropertyManager to select a location on a planer face for the center of the hole, pick the top of the part as shown.

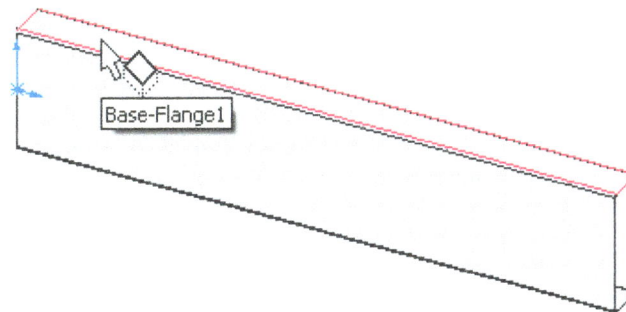

Base-Flange1

Check the **Link to thickness** check box.

In the **Hole** PropertyManager, enter '.75' for the **Hole Diameter**.

Click the green check mark button to accept the settings and create the hole.

In the FeatureManager design tree, slowly click two times (left click once, pause, left click again) on **Hole1** to select it.

Then, enter '**Rnd-.75**' for the new name and press **Enter**.

Since the hole is placed at the cursor location, you need to edit the sketch in order to locate the hole. In the FeatureManager design tree, right click on **Rnd-.75** and pick **Edit Sketch**.

Change the display view to **Normal To** the sketch by picking **Normal To** in the View list in the lower left corner of the graphics area or press **Ctrl-8**.

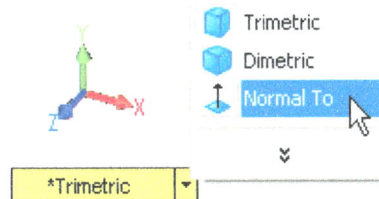

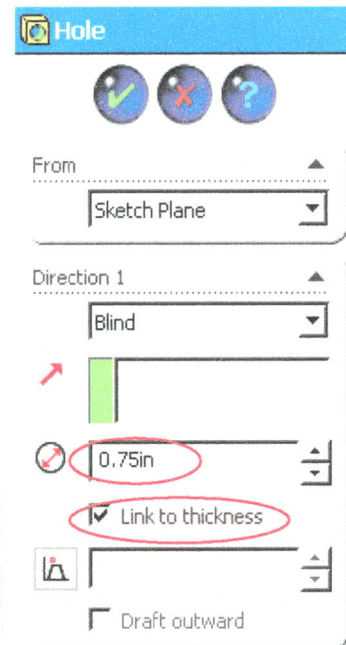

Hole

From
Sketch Plane

Direction 1
Blind

0.75in

Link to thickness

Draft outward

Trimetric
Dimetric
Normal To

*Trimetric

Now, zoom in closer to the right end of the part by clicking the **Zoom To Area** icon from the "View" toolbar. Make a box around the area that you want to zoom in to. Or scroll the middle mouse wheel, if you have one.

Add a '**1.125**' vertical dimension from the top of the part to the hole and a '**.875**' horizontal dimension from the right edge of the part to the hole as shown using the **Smart Dimension** icon in the CommandManager, or pull down the "Tools" menu and pick **Dimensions – Smart**.

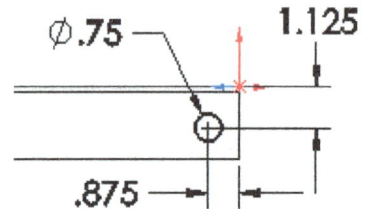

Press the **f** key on the keyboard to **Zoom to Fit** so you can see the entire edge flange and center it in the graphics area. (Keyboard shortcut key, **Zoom to Fit**: f)

Create a vertical centerline through the center of the part using the **Centerline** icon in the CommandManager, or pull down the "Tools" menu and pick **Sketch Entities – Centerline**.

Click the **Mirror Entities** icon from the CommandManager, or pull down the "Tools" menu and pick **Sketch Tools – Mirror**.

In the **Mirror** PropertyManager, under **Entities to mirror**, select the circle. Under **Mirror about** select the vertical centerline. You may have to select this line twice. The first time removes it from the **Entities to mirror** box if it was there.

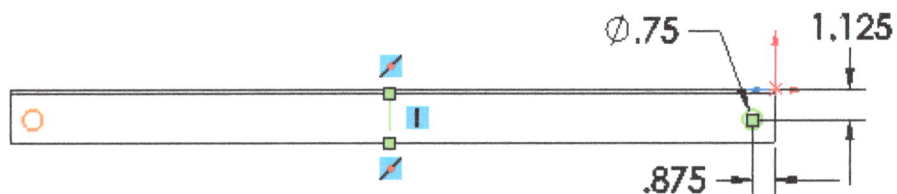

Click the green check mark button to accept the settings.

Exit the sketch by clicking the **Exit Sketch** icon in the CommandManager or in the upper right corner of the graphics area.

Click the **Sheet Metal** icon in the control area of the CommandManager. Then, click the **Simple Hole** icon in the toolbar, or pull down the "Insert" menu and pick **Features – Hole – Simple**.

When prompted in the PropertyManager to select a location on a planer face for the center of the hole, pick the front of the edge flange toward the right end as shown.

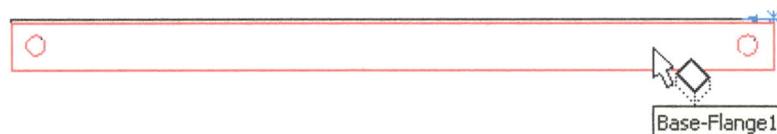

In the **Hole** PropertyManager, check the **Link to thickness** check box. Do not worry about the **Hole Diameter**. It should already be set to 0.75, but you will change the way it is done in just a minute.

Click the green check mark button to accept the settings and create the hole.

96   SolidWorks for the Sheet Metal Guy

In the FeatureManager design tree, slowly click two times (left click once, pause, left click again) on **Hole2** to select it.

Then, enter '**Rnd-.75 Seed**' for the new name and press **Enter**. This is the seed (starting) hole for the line pattern.

In the FeatureManager design tree, right click on **Rnd-.75 Seed** and pick **Edit Sketch**.

Delete the **.75** dimension. Next, select the circle (Make sure you select the circle, not the centerpoint of the circle). Then, hold down the **Ctrl** key and select the hole to the right which you created in the previous step.

In the **Properties** PropertyManager, click the **Equal** button.

Now, zoom in closer to the right end of the part by clicking the **Zoom To Area** icon from the "View" toolbar. Make a box around the area that you want to zoom in to. Or scroll the middle mouse wheel, if you have one.

Create a horizontal centerline through the center of the rightmost hole using the **Centerline** icon in the CommandManager, or pull down the "Tools" menu and pick **Sketch Entities – Centerline**.

Click the **Centerline** icon again to deselect it.

With the centerline selected (if it is green, it is selected), hold down the **Ctrl** key and select the centerpoint of the circle. (Make sure you select the centerpoint, not the circle).

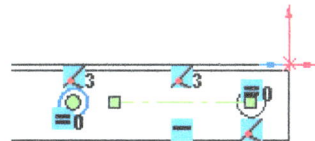

In the **Properties** PropertyManager, click the **Coincident** button.

Add a '**4.125**' horizontal dimension from the rightm ost hole to the circle as shown using the **Smart Dimension** icon in the CommandManager, or pull down the "Tools" menu and pick **Dimensions – Smart**.

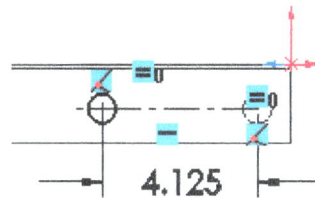

Exit the sketch by clicking the **Exit Sketch** icon in the CommandManager or in the upper right corner of the graphics area.

Create a Line of Holes

Click the **Features** icon in the control area of the CommandManager. Then, click the **Linear Pattern** icon from the CommandManager, or pull down the "Insert" menu and pick **Pattern/Mirror – Linear Pattern**.

Linear Pattern
Patterns features, faces, and bodies in one or two linear directions.

Press the **f** key on the keyboard to **Zoom to Fit** so you can see the entire edge flange and center it in the graphics area. (Keyboard shortcut key, **Zoom to Fit**: f)

Select the bottom horizontal edge of the part near the right end to set the direction of the pattern.

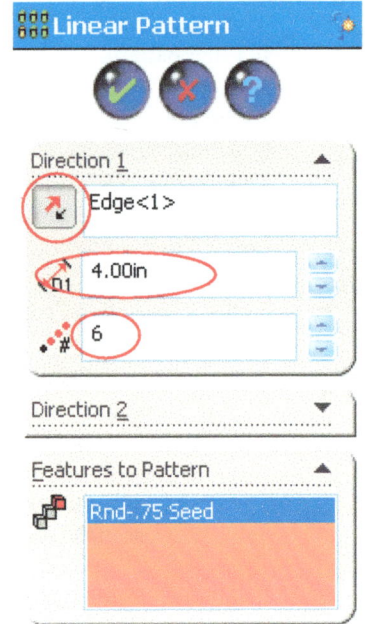

In the **Linear Pattern** PropertyManager, set the **Spacing** to '**4**'.

Set the **Number of Instances** to '**6**'.

Click in the **Features to Pattern** box.

In the flyout FeatureManager design tree, select **Rnd-.75 Seed**.

You should now see the yellow preview circles of the line pattern. If the line pattern is going to the right and off of the part, click the **Reverse Direction** button under **Direction 1**.

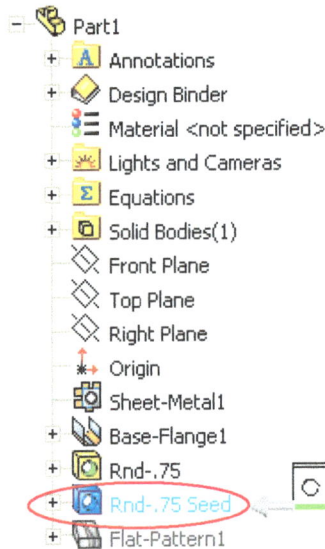

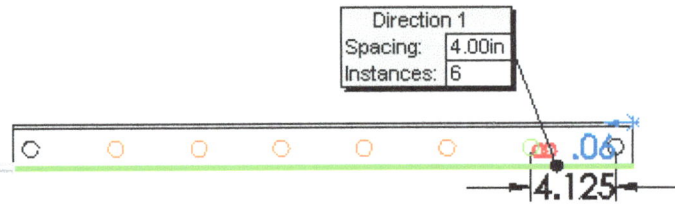

Click the green check mark button at the top of the **Linear Pattern** PropertyManager.

Change the display view to **Trimetric** by picking **Trimetric** in the View list in the lower left corner of the graphics area.

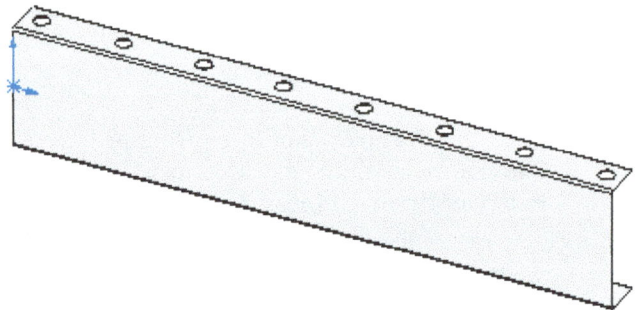

Save the Part

Click the **Save** icon in the "Standard" toolbar, or pick **Save** from the "File" pull down menu.

The **Save As** dialog box appears. In the **File name** box, type '**Damper Channel**' and click **Save**.

Prepare for the Design Table

To make it easier for you to know what the variables are in the Design Table, you can rename them. This allows you to assign a more meaningful name and hopefully makes it easier for you to understand the formulas.

In the FeatureManager design tree, right click on the **Annotations** folder and pick **Show Feature Dimensions** from the menu.

Locate the overall length 30.00 dimension and right click on it. Pick **Properties** from the menu. You may need to rotate the model in order to see and select some of these dimensions.

In the **Dimension Properties** dialog box, change the **Name** D2 to 'Length' and click **OK**.

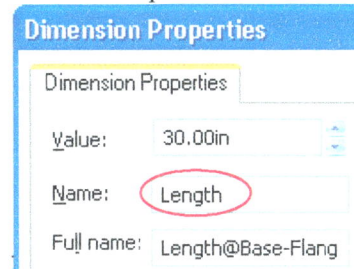

Dimension Properties

Dimension Properties

Value: 30.00in

Name: Length

Full name: Length@Base-Flang

Right click on the 2.00 dimension and pick **Properties** from the menu. In the **Dimension Properties** dialog box, change the **Name** D2 to 'Width' and click **OK**.

Right click on the .875 dimension and pick **Properties** from the menu. In the **Dimension Properties** dialog box, change the **Name** D3 to 'EndDim' and click **OK**.

Next to the .875 dimension is the 4.125 dimension. Right click on the 4.125 dimension and pick **Properties**. In the **Dimension Properties** dialog box, change the **Name** D1 to 'OddSpace' and click **OK**.

Next, change the name of the linear pattern dimensions by right clicking on the 4.00 dimension and pick **Properties**. In the **Dimension Properties** dialog box, change the **Name** D3 to 'Fl_Spacing', short for 'flange spacing', and click **OK**.

Then, right click on the 6 dimension and pick **Properties**. In the **Dimension Properties** dialog box, change the **Name** D1 to 'Fl_Instances' and click **OK**.

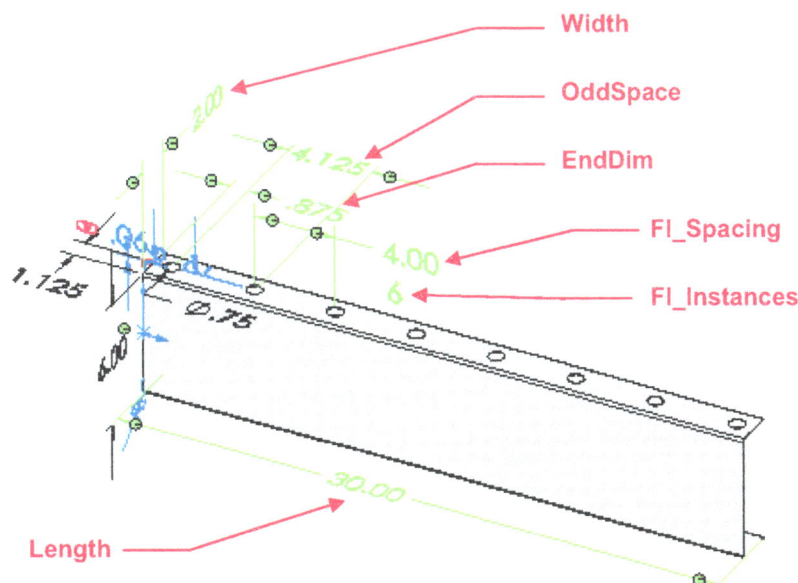

Create a Design Table

A design table allows you to specify parameters in an embedded Microsoft Excel worksheet. Formulas, "IF" logic, and other standard Excel features can be applied to the model.

Pull down the "Insert" menu and pick **Design Table**.

In the **Design Table** PropertyManager, under **Source**, click on **Blank** to insert a blank design table, allowing you fill to in the desired parameters.

Click the green check mark button at the top of the **Design Table** PropertyManager.

In the **Add Rows and Columns** dialog box, click the **OK** button.

The embedded Microsoft Excel worksheet opens inside of the graphics window. You may resize the table and drag the table to a different location using the border.

Click in cell A3 and type '**Default**' and press **Enter**.

Click in cell B2. In the graphics area, double click on the 30.00 dimension. If the table is covering over the dimensions, place the cursor on the edge of the table, press and hold the left mouse button, and drag the table to a new location.

If the mouse is off of the table when you do this, it will close the table. Should this happen, find the **Design Table** folder in the FeatureManager design tree and right click it. Pick **Edit table** and you're right back where you started.

The variable name, **Length@Base-Flange1** should appear in the cell B2 and the dimension value is automatically placed for you in cell B3.

Click in cell C2. In the graphics area, double click on the 2.00 dimension. **Width@Sketch2** should appear in the cell and the dimension is automatically placed for you in cell C3.

Click in cell D2. In the graphics area, double click on the 0.875 dimension. **EndDim@Sketch2** should appear in the cell and the dimension is automatically placed for you in cell D3.

Select cell E2, in the graphics area, double click on the 4.125 dimension. **OddSpace@Sketch3** should appear in the cell and the dimension is automatically placed for you in cell E3.

Select cell F2, in the graphics area, double click on the 4.00 dimension. **Fl_Spacing@LPattern1** should appear in the cell and the dimension is automatically placed for you in cell F3.

Select cell G2, in the graphics area, double click on the 6 dimension. **Fl_Instances@Lpattern1** should appear in the cell and the dimension is automatically placed for you in cell G3.

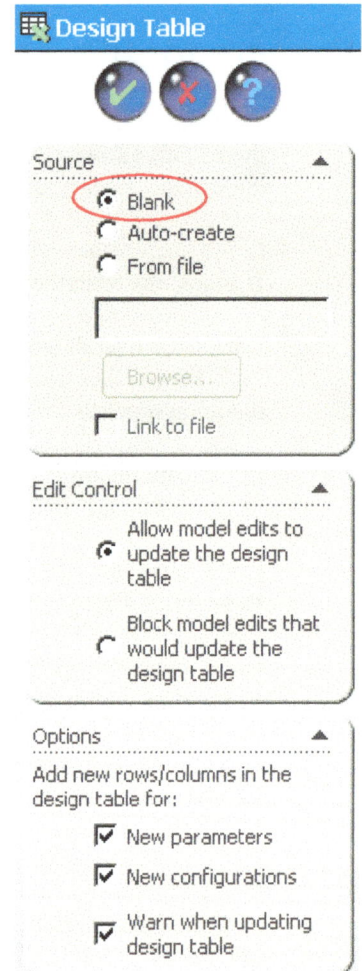

	A	B	C	D	E	F	G
1	Design Table for: Damper Channel						
2		Length@Base-Flange1	Width@Sketch1	EndDim@Sketch3	OddSpace@Sketch4	Fl_Spacing@LPattern1	Fl_Instances@LPattern1
3	Default	30	2	0.875	4.125	4	6
4							
5							

Sheet1

Using the scroll arrow in the bottom right-hand corner of the Design Table, scroll over so that you can see up to cell L3.

Click in cell K1 and type 'True Instances'. This goes in the first row, because the second row is reserved for variable names. Also, you must place this in a column to the right of your variables or the table will not update all of the variables. You skipped a few columns to save room for more variables which will be added later.

The number of true instances is calculated by subtracting two times the **EndDim** from the overall length and dividing by the spacing distance. You call it the "true" value because it may need to be adjusted if the **OddSpace** value is too small.

Click in cell K3 and type '=INT((B3-D3*2)/F3)'. The value **7** should appear when you press the enter key on the keyboard.

Click in cell L1 and type 'True OddSpace'.

This is the "true" calculated **OddSpace** value. However, you will build into the formulas that it cannot be less than 70% of the spacing value and that may cause the **OddSpace** value you use to be different than this amount.

Click in cell L3 and type '=(B3-D3*2-(K3-1)*F3)/2'. The value **2.125** should appear when you press the enter key on the keyboard.

Go back and click in cell G3 and type '=IF(L3<0.7*F3,K3-1,K3)'. The value should remain **6** when you press the enter key on the keyboard. The "IF" statement is testing whether the calculated OddSpace is less than 70% of the Fl_Spacing. If it is, it subtracts 1 from the number of spaces (**Fl_Instances**).

Click in cell E3 and type '=(B3-D3*2-(G3-1)*F3)/2'. The value should remain **4.125** when you press the **Enter** key on the keyboard. The Odd Space is equal to the Length minus 2 times the Width minus the number of steps (instances – 1) times the spacing and divide it all by two.

	E	F	G	H	I	J	K	L
1							True Instances	True OddSpace
2	OddSpace@Sketch4	Fl_Spacing@LPattern1	Fl_Instances@LPattern1					
3	4.125	4	6				7	2.125
4								
5								

Sheet1

Click anywhere in the graphics area to exit the Design Table and apply the changes.

The dimensions that you included in the Design Table appear in magenta color. In the FeatureManager design tree, right click on the **Annotations** folder and pick **Show Feature Dimensions** from the menu. This will turn off the dimensions and clean up the graphics area a little.

Mirror the Holes to the Lower Flange

Click the **Features** icon in the control area of the CommandManager. Then, click the **Mirror** icon from the toolbar, or pull down the "Insert" menu and pick **Pattern/Mirror – Mirror**.

In the flyout FeatureManager design tree, select **Top Plane** for the **Mirror Face/Plane**.

In the **Mirror** PropertyManager, click in the **Features to Mirror** box. Select **Rnd-.75**, **Rnd-.75 Seed**, and **Lpattern1** in the flyout FeatureManager design tree.

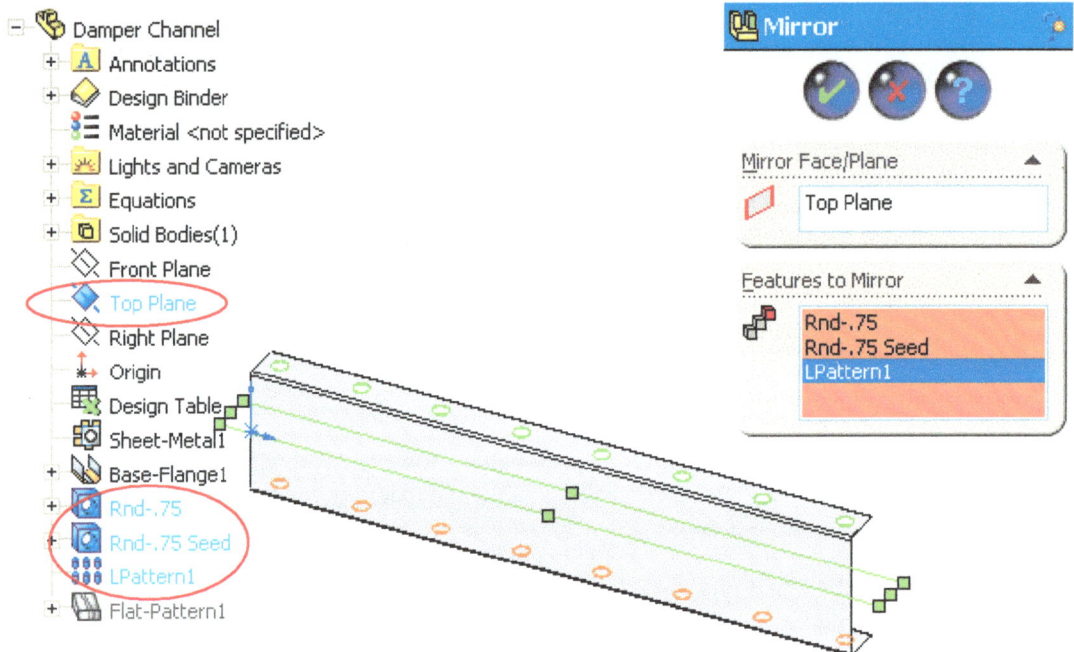

Click the green check mark button at the top of the **Mirror** PropertyManager.

Create the Shaft Holes

Click the **Sheet Metal** icon in the control area of the CommandManager. Then, click the **Simple Hole** icon in the toolbar, or pull down the "Insert" menu and pick **Features – Hole – Simple**.

Pick the front of the part where shown.

In the **Hole** PropertyManager, check the **Link to thickness** check box.

For the **Hole Diameter**, enter '**1.8125**'.

Click the green check mark button to create the hole.

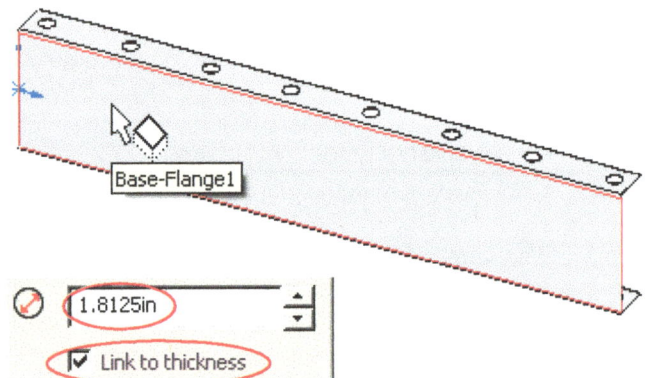

In the Feature Manager design tree, slowly click two times (left click once, pause, left click again) on **Hole3** to select it. Then, type '**Rnd-1.8125**' for the new name and press **Enter**.

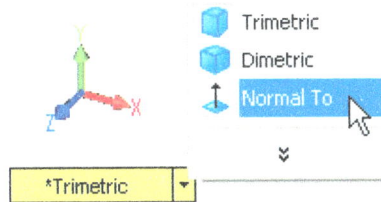

In the FeatureManager design tree, right click on **Rnd-1.8125** and pick **Edit Sketch**.

Change the display view to **Normal To** the sketch by picking **Normal To** in the View list in the lower left corner of the graphics area.

Ctrl select the centerpoint of the circle and the origin.

In the **Properties** PropertyManager, click the **Horizontal** button.

Add a '**6.25**' horizontal dimension from the left edge of the part to the hole as shown using the **Smart Dimension** icon in the CommandManager, or pull down the "Tools" menu and pick **Dimensions – Smart**.

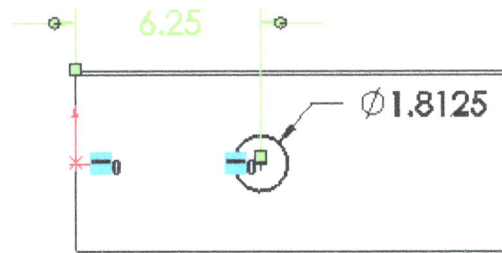

You can rename the dimension when you create it. Right click on the 6.25 dimensions and pick **Properties**.

In the **Dimension Properties** dialog box, change the **Name D2** to '**EndDistance**' and click **OK**.

Exit the sketch by clicking the **Exit Sketch** icon in the CommandManager or in the upper right corner of the graphics area.

Create the Seed Hole for the Bolt Hole Circle

Click the **Sheet Metal** icon in the control area of the CommandManager. Then, click the **Simple Hole** icon in the toolbar, or pull down the "Insert" menu and pick **Features – Hole – Simple**.

When prompted in the PropertyManager to select a location on a planer face for the center of the hole, pick the front of the part just above the 1.8125 circle as shown.

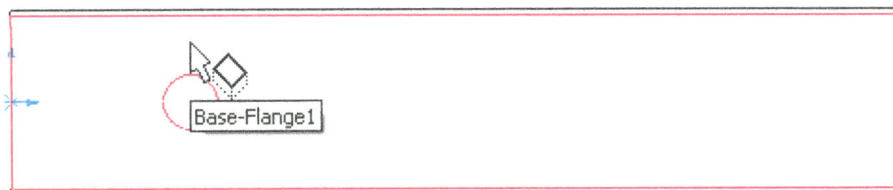

In the **Hole** PropertyManager, check the **Link to thickness** check box and enter '**.3125**' (5/16) for the **Hole Diameter**.

Click the green check mark button to create the hole.

In the FeatureManager design tree, slowly click two times (left click once, pause, left click again) on **Hole4** to select it. Then, type '**Rnd-.3125**' for the new name and press **Enter**.

In the FeatureManager design tree, right click on **Rnd-.3125** and pick **Edit Sketch**.

Create a vertical centerline through the center of the existing 1.8125 diameter hole using the **Centerline** icon in the CommandManager, or pull down the "Tools" menu and pick **Sketch Entities – Centerline**.

Click the **Centerline** icon again to deselect it.

With the centerline selected, hold down the **Ctrl** key and select the centerpoint of the 0.3125 diameter circle. (Make sure you select the centerpoint of the circle, not the circle).

In the **Properties** PropertyManager, click the **Coincident** button.

Add a '**1.75**' vertical dimension from the hole to the circle as shown using the **Smart Dimension** icon in the CommandManager, or pull down the "Tools" menu and pick **Dimensions – Smart**.

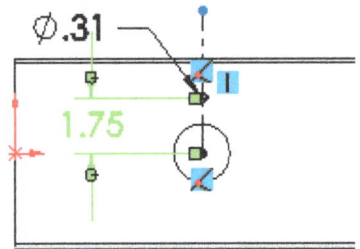

Right click on the 1.75 dimension and pick **Properties**. In the **Dimension Properties** dialog box, change the **Name** D2 to '**BHC_Radius**' and click **OK**.

Exit the sketch by clicking the **Exit Sketch** icon in the CommandManager or in the upper right corner of the graphics area.

Create the Bolt Hole Circle

Click the **Features** icon in the control area of the CommandManager. Then, click the **Circular Pattern** icon from the CommandManager, or pull down the "Insert" menu and pick **Pattern/Mirror – Circular Pattern**.

SolidWorks requires that you select an axis, model edge, or an angular dimension in the graphics area to rotate the feature around the axis.

In order to do this, pull down the "View" menu and pick **Temporary Axes** to display the temporary axis of all of the round holes.

In the graphics area, select the temporary axis of **Rnd-1.8125** as shown below. The cursor will change indicating that an axis is selected. You may need to zoom in to select the temporary axis.

In the **Circular Pattern** PropertyManager, under **Parameters**, check the **Equal spacing** check box.

Enter '**360**' for the **Total Angle**.

Enter '**6**' for the **Number of Instances**.

Click in the **Features to Pattern** box. In the flyout FeatureManager design tree, select **Rnd-.3125**.

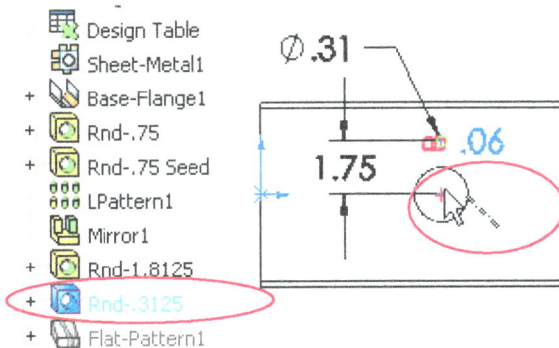

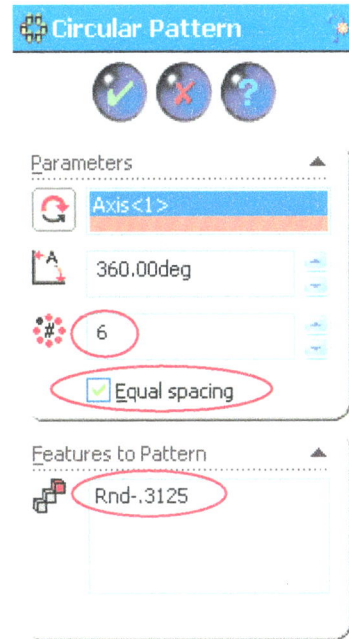

Click the green check mark button at the top of the **Circular Pattern** PropertyManager.

Create a Line of Bolt Hole Circles

Click the **Features** icon in the control area of the CommandManager. Then, click the **Linear Pattern** icon from the CommandManager, or pull down the "Insert" menu and pick **Pattern/Mirror – Linear Pattern**.

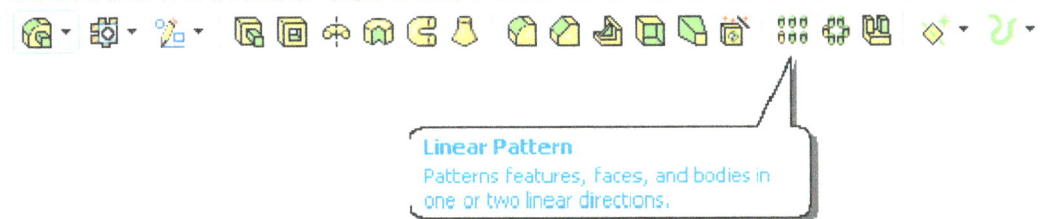

Linear Pattern
Patterns features, faces, and bodies in one or two linear directions.

Select the bottom edge of the part below the bolt hole circle to set the direction of the pattern.

In the **Linear Pattern** PropertyManager, set the **Spacing** to '**8**'.

Set the **Number of Instances** to '**3**'.

Click in the **Features to Pattern** box.

In the flyout FeatureManager design tree, select **CircularPattern1** and **Rnd-1.8125**.

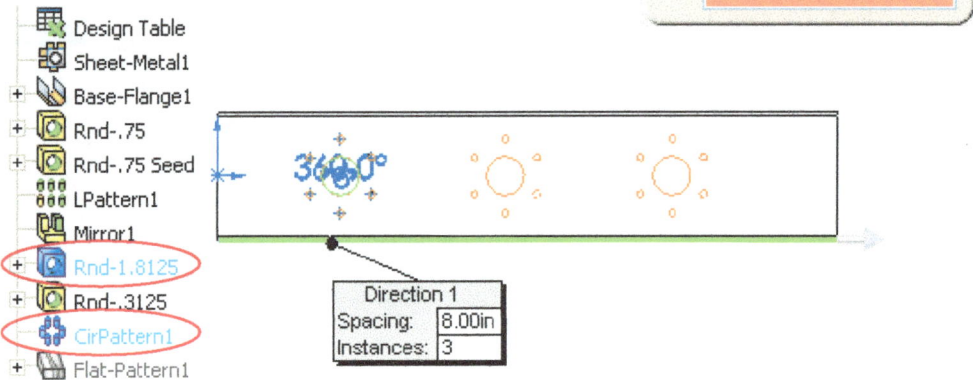

Click the green check mark button at the top of the **Linear Pattern** PropertyManager.

Pull down the "View" menu and pick **Temporary Axes** to hide the display of the temporary axes.

Change the display view to **Trimetric** by picking **Trimetric** in the View list in the lower left corner of the graphics area.

Change the Dimension Names

In the FeatureManager design tree, right click on the **Annotations** folder and pick **Show Feature Dimensions** from the menu.

Now to temporarily clear off some of the dimension mess, in the FeatureManager design tree, right click on **Rnd-.75** and pick **Suppress** from the menu.

This removes the feature and its related features and dimensions from the display, making it much easier to see the dimensions for the line of bolt hole circles.

Now, locate the linear pattern spacing and number of instances dimensions.

Right click on the 8.00 dimension and pick **Properties** from the menu.

In the **Dimension Properties** dialog box, change the **Name** D3 to 'Sh_Spacing' and click **OK**.

Right click on the 3 dimension and pick **Properties**.

In the **Dimension Properties** dialog box, change the **Name** D1 to 'Sh_Instances' and click **OK**.

Press the **Escape** key on the keyboard to close any active commands.

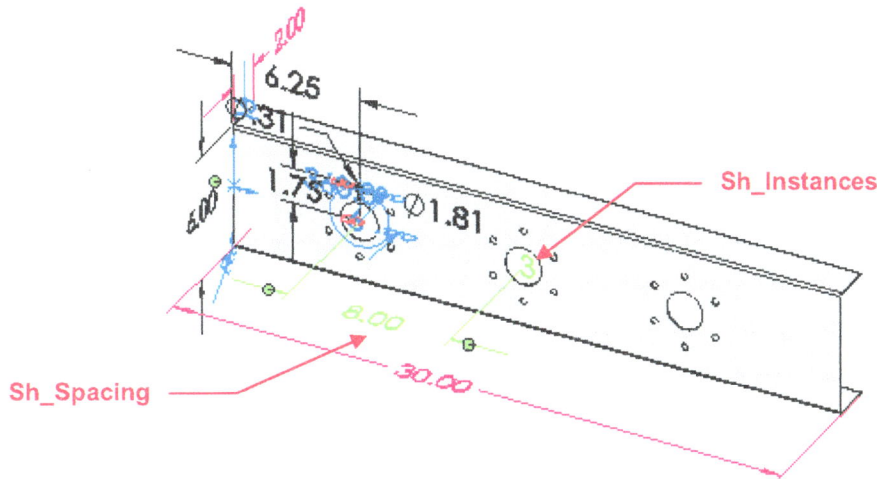

Sh_Instances

Sh_Spacing

Add Equations by Editing the Design Table

In the FeatureManager design tree, right click on **Design Table**, and pick **Edit Table**.

Click **OK** in the **Add Rows and Columns** dialog box.

With cell H2 selected, in the graphics area, double click on the 6.25 end distance dimension. **EndDistance@Sketch5** should appear in the cell and the dimension is automatically placed for you in cell H3. Do not worry if the sketch number is different.

With cell I2 selected, in the graphics area, double click on the 8.00 spacing dimension. **Spacing@Lpattern2** should appear in the cell and the dimension is automatically placed for you in cell I3.

With cell J2 selected, in the graphics area, double click on the 3 instances dimension. **Instances@Lpattern2** should appear in the cell and the dimension is automatically placed for you in cell J3.

Click in cell H3 and type '**=C3+I3/2**' and press **Enter**.

Click in cell I3 and type '**=(B3-(2*C3))/J3**' and press **Enter**.

Click in cell J3 and type '**=INT((B3-(C3*2))/M3)**' and press **Enter**. Don't worry if the cells show #DIV/0!. This will be fixed in the next step.

Columns K and L relate to the flange holes, so skip over to column M.

Click in cell M1 and type '**Ideal Blade Width**'.

Click in cell M3 and type '**7**' and press **Enter**.

	H	I	J	K	L	M	N
1				True Instances	True OddSpace	Ideal Blade Width	
2	EndDistance@Sketch5	Sh_Spacing@LPattern2	Sh_Instances@LPattern2				
3	6.333333333	8.666666667	3	7	2.125	7	
4							

⊮ ◀ ▶ ⊮ \ **Sheet1** /

Click anywhere in the graphics area or on the green check mark in the upper right hand corner of the graphics area to exit the Design Table and apply the changes.

Press the **Escape** key on the keyboard to close any active commands. The dimensions that you used in the Design Table appear in the magenta color.

To make the part fully parametric, you should have included all of the part dimensions in the Design Table. You chose at this time to keep it short. A further discussion of Design Tables will be included in Course 4.

In the FeatureManager design tree, click on **Rnd-.75**. Hold down the **Shift** key and click on **Mirror1**. With **Rnd-.75**, **Rnd-.75 Seed**, **LPattern1**, and **Mirror1** highlighted, right click on **Rnd-.75** and pick **Unsuppress** from the menu.

In the FeatureManager design tree, right click on the **Annotations** folder and pick **Show Feature Dimensions** from the menu. This turns off the dimensions so you only see the part.

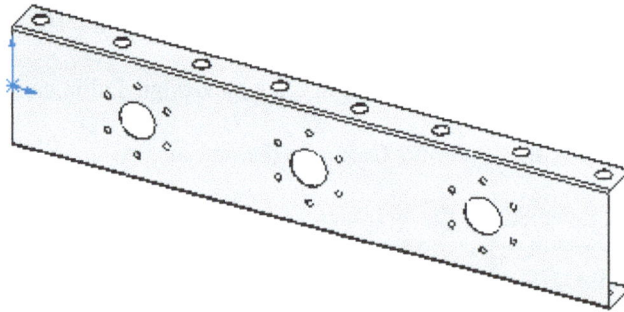

Click the **Save** icon in the "Standard" toolbar, or pick **Save** from the "File" pull down menu.

Now It Is Play Time!

In the FeatureManager design tree, right click on **Design Table**, and pick **Edit Table**.

Change the value in B3 from 30 to '**24**'.

Click anywhere in the graphics area or on the green check mark in the upper right hand corner of the graphics area to exit the Design Table.

The Design Table closes and the part is regenerated using the new dimensions. Everything updates based on the formulas you created in the Design Table.

Try changing the length (B3 in the Design Table) to some other values. Take a closer look at the **OddSpacing** value in the table to see that it is updated correctly.

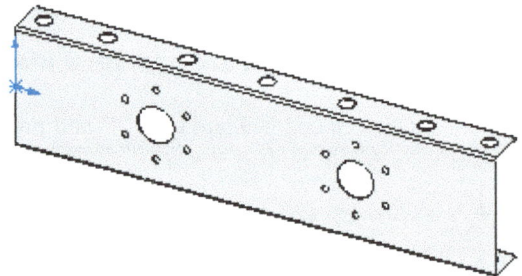

Chapter 8

Door Panel

The Door Panel is a very common part. Start with a grid pattern of Louvers for ventilation. Then, add holes to mount the handle, a Double D shape, and a couple of rounds. Use relations to keep them together as a group. A rectangle cutout on the side of the part allows the latch to pass through. This too must be related to the handle geometry. Do not forget the hinges.

A large cutout is placed at the bottom and a fill pattern set below. These are then tied together so that they will update each other. You should play with the variables in the Fill Pattern dialog box. It is interesting what capabilities exist here, and with some creativity, they can help you solve some complex hole problems.

Create the Base Flange

Begin a new **Part** document by clicking the **New** icon in the "Standard" toolbar, or pull down the "File" menu and pick **New**.

Create a base flange by clicking the **Sheet Metal** icon in the control area of the CommandManager. Then, click the **Base-Flange/Tab** icon from the toolbar, or pull down the "Insert" menu and pick **Sheet Metal – Base Flange**.

Select the **Front** plane when prompted to select a plane on which to sketch the feature cross-section.

Create a rectangle with the origin inside the rectangle using the **Rectangle** icon in the CommandManager, or pull down the "Tools" menu and pick **Sketch Entities – Rectangle**.

Create a construction line diagonally across the rectangle by clicking the **Centerline** icon in the CommandManager, or pull down the "Tools" menu and pick **Sketch Entities – Centerline**.

Select the top left corner and then the bottom right corner of the rectangle that you just created to create the centerline as shown below.

Right click in the graphics area and pick **Select** from the menu, or press the **Escape** key.

Select the diagonal line. Then, hold down the **Ctrl** key and select the origin.

In the **Properties** PropertyManager, under **Add Relations**, click the **Midpoint** button.

Add a '**12**' horizontal dimension to the bottom line and a '**18**' vertical dimension to the left vertical line using the **Smart Dimension** icon in the CommandManager, or pull down the "Tools" menu and pick **Dimensions – Smart**.

Exit the sketch by clicking the **Exit Sketch** icon in the CommandManager or in the upper right corner of the graphics area.

In the **Base Flange** PropertyManager under **Sheet Metal Parameters**, set the **Thickness** to '**.048**'. Make sure that the **Reverse direction** check box is not checked.

Click the green check mark button at the top of the **Base Flange** PropertyManager to accept the settings and create the part.

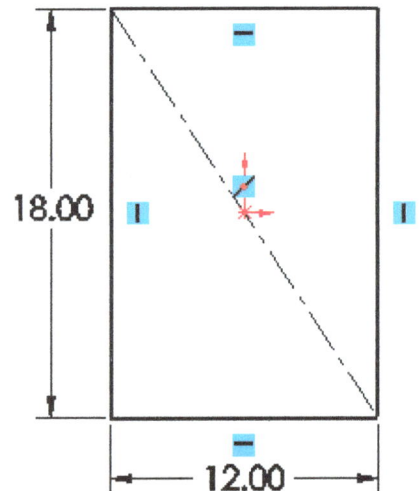

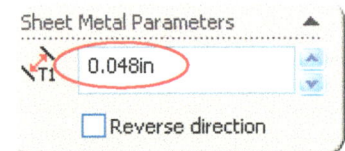

Insert a Punching Shape

With the Task Pane shown, click the **Design Library** icon in the Task Pane to expand the **Design Library** tab.

Design Library
Click to display this task pane tab.

In the **SheetMetalGuy** Design Library, click on your **Forming Tools** folder to display the folder contents in the lower pane of the **Design Library** tab.

Drag **2 x .375 louver** from the lower pane of the **Design Library** tab to the top left of the part (but do not release the mouse button yet). Press the **Tab** key to change the tool to be up instead of down as shown. Release the mouse button to place the tool.

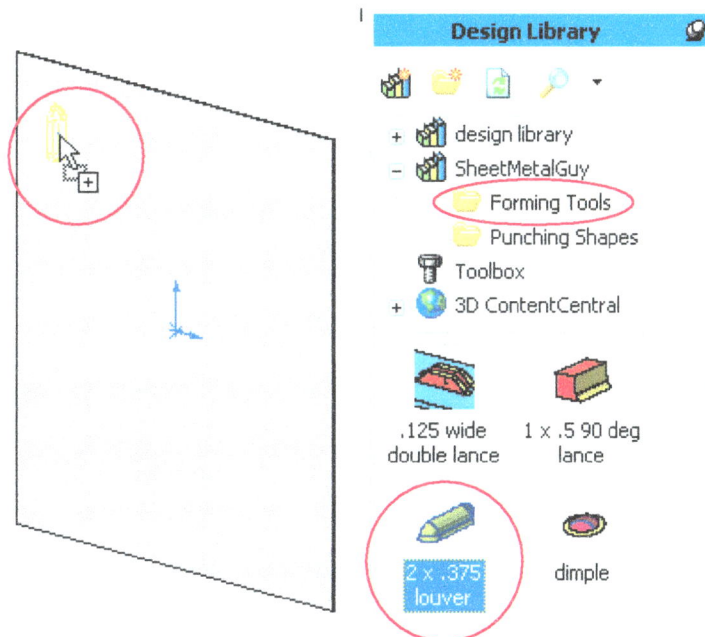

Pull down the "Tools" menu and pick **Sketch Tools – Modify**.

In the **Modify Sketch** dialog box, under **Rotate**, enter '**-90**' and press the **Enter** key. The sketch will rotate -90 degrees. You may also right click and spin the sketch in the graphics area.

In the **Modify Sketch** dialog box, click the **Close** button.

⟡ Click the **Smart Dimension** icon in the CommandManager, or pull down the "Tools" menu and pick **Dimensions – Smart**.

Add a '**3**' horizontal dimension from the part origin to the vertical centerline of the louver. Then, add a '**2**' vertical dimension from the top edge of the part to the horizontal centerline of the louver as shown.

In the **Position form feature** dialog box, click **Finish**.

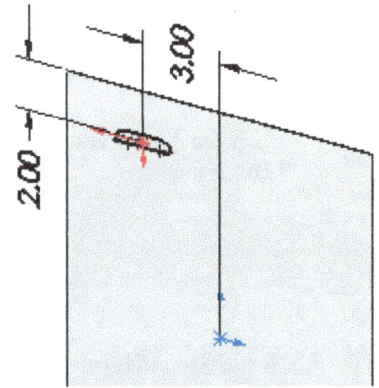

Create a Grid Pattern

⊞ Click the **Features** icon in the control area of the CommandManager. Then, click the **Linear Pattern** icon from the CommandManager, or pull down the "Insert" menu and pick **Pattern/Mirror – Linear Pattern**.

In the graphics area, select the left half of the top edge of the part to set the **Direction 1** of the pattern as shown.

In the **Linear Pattern** PropertyManager, under **Direction 1**, set the **Spacing** to '**3**' and the **Number of Instances** to '**3**'.

Under **Direction 2**, click in the **Pattern Direction** box.

In the graphics area, select the top half of the left edge of the part to set the **Direction 2** of the pattern.

Under **Direction 2**, set the **Spacing** to '**1**' and the **Number of Instances** to '**5**'.

Click in the **Features to Pattern** box and make sure **2 x .375 louver1** is selected. If not, in the flyout FeatureManager design tree, select **2 x .375 louver1**.

✓ Click the green check mark button at the top of the **Linear Pattern** PropertyManager.

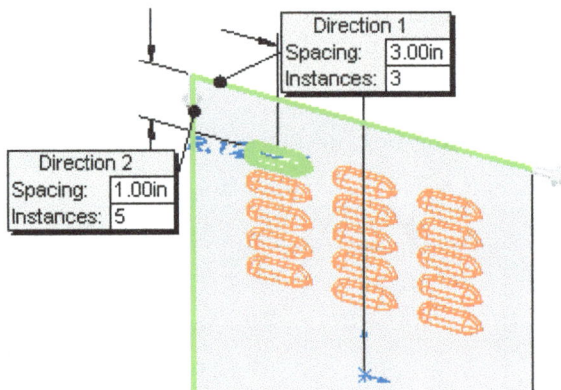

Add an Edge Flange

Click the **Sheet Metal** icon in the control area of the CommandManager. Then, click the **Edge Flange** icon in the CommandManager, or pull down the "Insert" menu and pick **Sheet Metal – Edge Flange**.

Select the right edge of the part and move the cursor to the right and click to set the direction of the flange. Then, select the other three sides as shown.

In the **Edge-Flange** PropertyManager, set the **Flange Length** to **Blind** and the **Length** to '**1**'.

Click the **Outer Virtual Sharp** button and the **Material Inside Flange Position** button.

Click the green check mark button at the top of the **Edge-Flange** PropertyManager to accept the settings and create the flange.

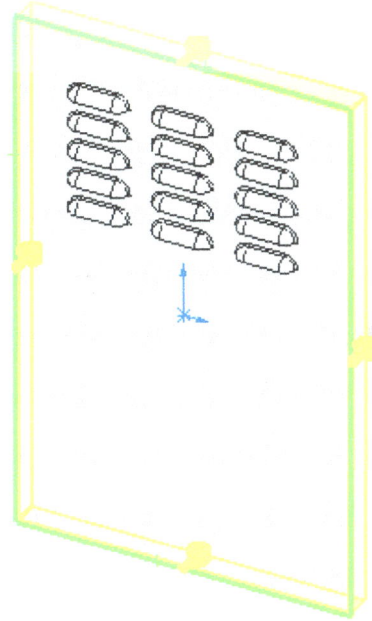

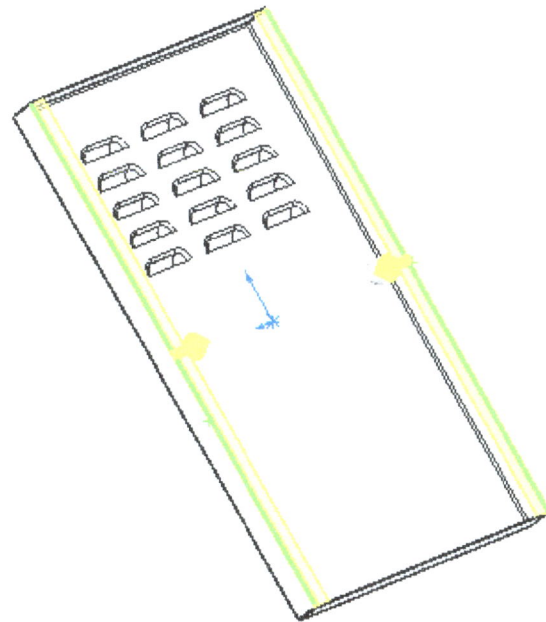

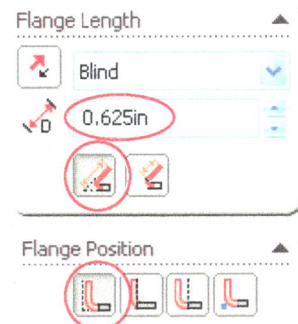

Add Another Edge Flange

Press the **Left arrow** key on the keyboard seven times to rotate the part to the back side.

Click the **Edge Flange** icon in the CommandManager again, or pull down the "Insert" menu and pick **Sheet Metal – Edge Flange**.

Select the long side edge flange that you just created and move the cursor in towards the middle of the part and click to set the direction of the flange. Then, select the other side flange as shown.

In the **Edge-Flange** PropertyManager, set the **Flange Length** to **Blind** and the **Length** to '**.625**'.

Click the **Outer Virtual Sharp** button and the **Material Inside Flange Position** button.

Click the green check mark button at the top of the **Edge-Flange** PropertyManager to accept the settings and create the flange.

Create the Handle Opening

Click the **Extruded Cut** icon from the CommandManager or pull down the "Insert" menu and pick **Cut – Extrude**.

In the bottom left corner of your graphics area, change the View orientation by clicking the pull down arrow and picking **Front**.

Select the front of the part for the plane to sketch on.

Create a circle as shown using the **Circle** icon in the CommandManager, or pull down the "Tools" menu and pick **Sketch Entities – Circle**.

Then, create a vertical line in the left half of the circle and another in the right half as shown using the **Line** icon in the CommandManager, or pull down the "Tools" menu and pick **Sketch Entities – Line**.

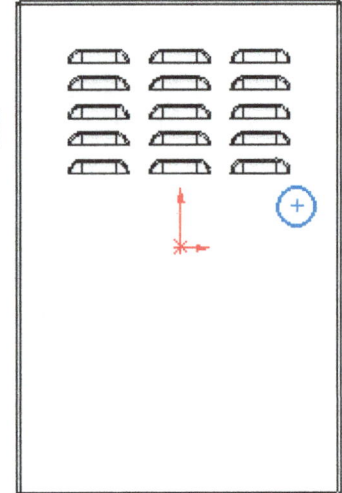

In the graphics area, right click and pick **Select** or press the **Escape** key.

Select the left vertical line. Then, hold down the **Ctrl** key and select the right vertical line.

In the **Properties** PropertyManager, click on the **Equal** button.

Next, click the **Trim Entities** icon or pull down the "Tools" menu and pick **Sketch Tools - Trim Entities**.

In the **Trim** PropertyManager, make sure that the **Trim to closest** button is depressed, and select the left section of the circle and right section of the circle. This forms the Double D of the hole.

Click the **Smart Dimension** icon in the CommandManager, or pull down the "Tools" menu and pick **Dimensions – Smart**.

Add a '.75' dimension between the vertical lines and a '.5' radius dimension to the circle.

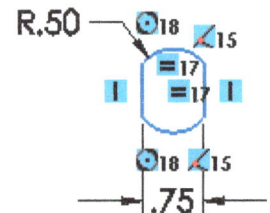

To make the circle dimension a diameter dimension, right click on the R.50 dimension and pick **Properties**.

In the **Dimension Properties** dialog box, check the **Diameter dimension** check box.

Click the **OK** button.

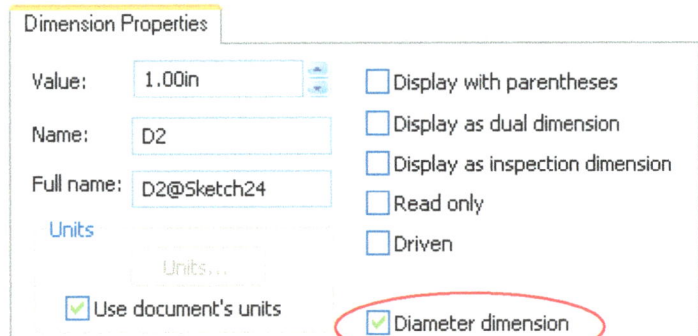

SolidWorks for the Sheet Metal Guy

Create a '.3125' diameter circle to the above the shape as shown below using the **Circle** icon in the CommandManager, or pull down the "Tools" menu and pick **Sketch Entities – Circle**.

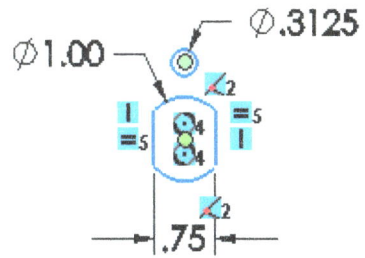

Press the **Escape** key, and then select the centerpoint of the circle. Hold down the **Ctrl** key and select the centerpoint of the Double D.

In the **Properties** PropertyManager, click on the **Vertical** button.

Create a horizontal centerline through the middle of the shape as shown using the **Centerline** icon in the CommandManager, or pull down the "Tools" menu and pick **Sketch Entities – Centerline**.

Click the **Mirror Entities** icon from the CommandManager, or pull down the "Tools" menu and pick **Sketch Tools – Mirror**.

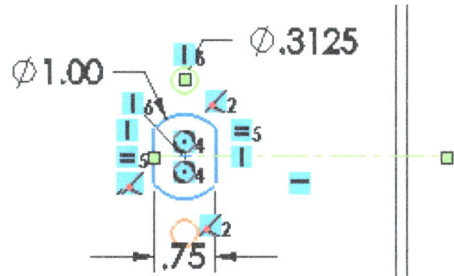

In the **Mirror** PropertyManager, under **Entities to mirror**, select the .3125 circle. Under **Mirror about** select the horizontal centerline.

Click the green check mark button at the top of the **Mirror** PropertyManager.

Click the **Smart Dimension** icon in the CommandManager, or pull down the "Tools" menu and pick **Dimensions – Smart**.

Add a '**1.25**' dimension from the center of the circle to the right edge of the part. Add a '**.875**' dimension from the bottom circle to the horizontal centerline. Finally add an '**8**' dimension from the centerline to the top edge of the part.

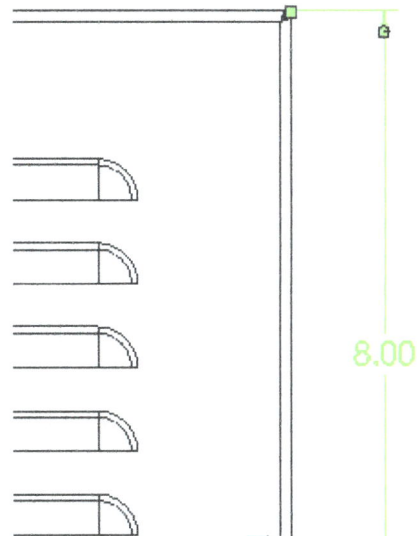

Exit the sketch by clicking the **Exit Sketch** icon in the CommandManager or in the upper right corner of the graphics area.

In the **Cut-Extrude** PropertyManager, check the **Link to thickness** check box and then click the green check mark button at the top of the **Cut-Extrude** PropertyManager.

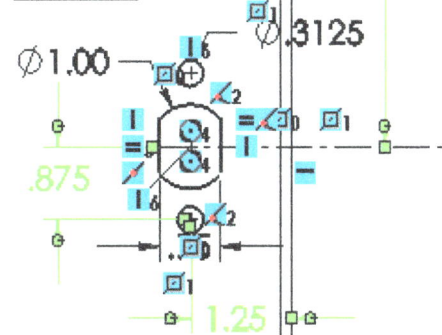

Create Holes on the Side

Click the **Sheet Metal** icon in the control area of the CommandManager. Then, click the **Extruded Cut** icon or pull down the "Insert" menu and pick **Cut – Extrude**.

In the bottom left corner of your graphics area, change the View orientation by clicking the pull down arrow and picking **Left**, or press **Ctrl-3**.

Select the left side of the part for the plane to sketch on.

Create two circles on the right side of the part using the **Circle** icon in the CommandManager, or pull down the "Tools" menu and pick **Sketch Entities – Circle**.

Press the **Escape** key, and then select the centerpoint of the top circle. Hold down the **Ctrl** key and select the centerpoint of the bottom circle.

In the **Properties** PropertyManager, click on the **Vertical** button.

Select the top circle, and then, hold down the **Ctrl** key and select the bottom circle. In the **Properties** PropertyManager, click on the **Equal** button.

Click the **Smart Dimension** icon in the CommandManager, or pull down the "Tools" menu and pick **Dimensions – Smart**.

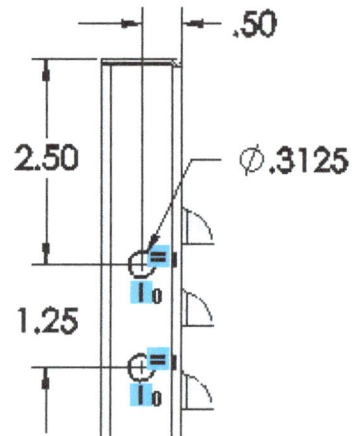

Add a '**.3125**' diameter dimension to the top circle and a '**.5**' horizontal dimension from the .3125 circle to the front of the part. Then, add a '**2.5**' vertical dimension from the top of the part to the .3125 circle, followed by a '**1.25**' vertical dimension between the two circles as shown above.

Create a horizontal centerline line through the origin as shown below using the **Centerline** icon in the CommandManager, or pull down the "Tools" menu and pick **Sketch Entities – Centerline**.

Click the **Mirror Entities** icon from the CommandManager, or pull down the "Tools" menu and pick **Sketch Tools – Mirror**.

In the **Mirror** PropertyManager, under **Entities to mirror**, select the two circles. Under **Mirror about** select the horizontal centerline. Then, at the top of the **Mirror** PropertyManager, click the green check mark button.

Exit the sketch by clicking the **Exit Sketch** icon in the CommandManager or in the upper right corner of the graphics area.

In the **Cut-Extrude** PropertyManager, check the **Link to thickness** check box and then click the green check mark button at the top of the **Cut-Extrude** PropertyManager.

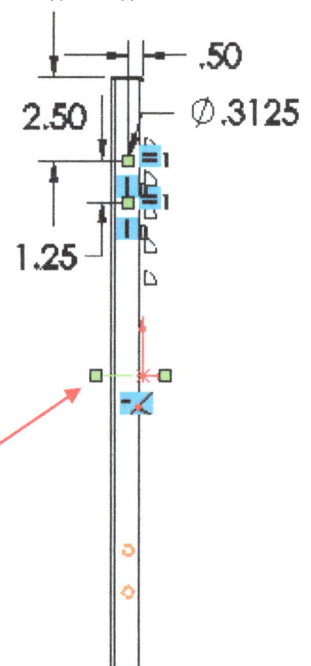

Create the Latch Opening

Click the **Sheet Metal** icon in the control area of the CommandManager. Then, click the **Extruded Cut** icon or pull down the "Insert" menu and pick **Cut – Extrude**.

In the bottom left corner of your graphics area, change the View orientation by clicking the pull down arrow and picking **Right**, or press **Ctrl-4**.

Select the right side of the part for the plane to sketch on.

Press the **Right arrow** key twice to rotate the part.

To make sure that if the latch opening ever moves, the lock opening must move also. A simple relation will accomplish this.

To do this, select the right vertical edge line of the latch opening as shown.

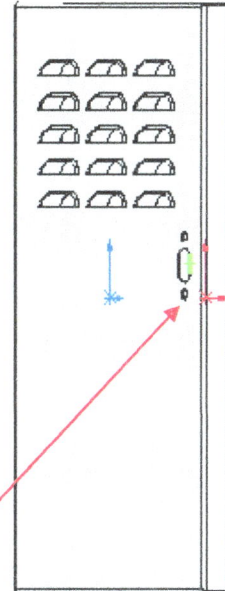

Click on the **Convert Entities** icon in the CommandManager, or pull down the "Tools" and pick **Sketch Tools – Convert Entities**.

Select the new line created.

In the **Line Properties** PropertyManager, under **Options**, check the **For construction** check box.

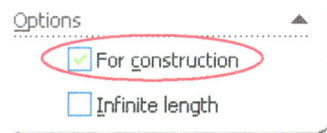

Press the **Left arrow** key twice to rotate the part back to the **Right** view.

Add a horizontal centerline line at the midpoint of the sketch line, adding an automatic relation, as shown using the **Centerline** icon in the CommandManager, or pull down the "Tools" menu and pick **Sketch Entities – Centerline**.

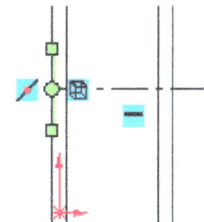

Create a rectangle as shown using the **Rectangle** icon in the CommandManager, or pull down the "Tools" menu and pick **Sketch Entities – Rectangle**.

Press the **Escape** key to deselect the command.

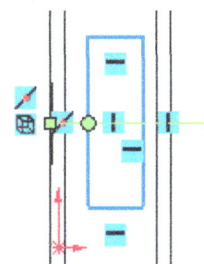

Right click on the left vertical line of the rectangle and pick **Select Midpoint**. Then, hold down the **Ctrl** key and select the centerline. In the **Properties** PropertyManager, click the **Coincident** button.

Create a vertical centerline connecting the midpoints of the horizontal lines of the rectangle using the **Centerline** icon in the CommandManager, or pull down the "Tools" menu and pick **Sketch Entities – Centerline**.

Click the **Smart Dimension** icon in the CommandManager, or pull down the "Tools" menu and pick **Dimensions – Smart**.

Add a '**.5**' horizontal dimension from the front of the part to the vertical centerline. Then, add a '**.25**' horizontal dimension to the rectangle, followed by a '**.875**' vertical dimension.

Exit the sketch by clicking the **Exit Sketch** icon in the CommandManager or in the upper right corner of the graphics area.

In the **Cut-Extrude** PropertyManager, check the **Link to thickness** check box and then click the green check mark button.

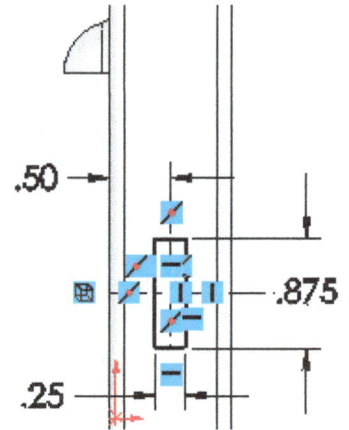

Create an Extruded Cut

Click the **Sheet Metal** icon in the control area of the CommandManager. Then, click the **Extruded Cut** icon or pull down the "Insert" menu and pick **Cut – Extrude**.

In the bottom left corner of your graphics area, change the View orientation by clicking the pull down arrow and picking **Front**.

Select the front of the part for the plane to sketch on.

Create a rectangle on the front of the part using the **Rectangle** icon in the CommandManager, or pull down the "Tools" menu and pick **Sketch Entities – Rectangle**.

Press the **Escape** key to deselect the command.

Right click on the top horizontal line of the rectangle and pick **Select Midpoint**. Then, hold down the **Ctrl** key and select the origin. In the **Properties** PropertyManager, click the **Vertical** button.

Add the dimensions as shown using the **Smart Dimension** icon in the CommandManager, or pull down the "Tools" menu and pick **Dimensions – Smart**.

Create a circle outside each corner of the rectangle using the **Circle** icon in the CommandManager, or pull down the "Tools" menu and pick **Sketch Entities – Circle**.

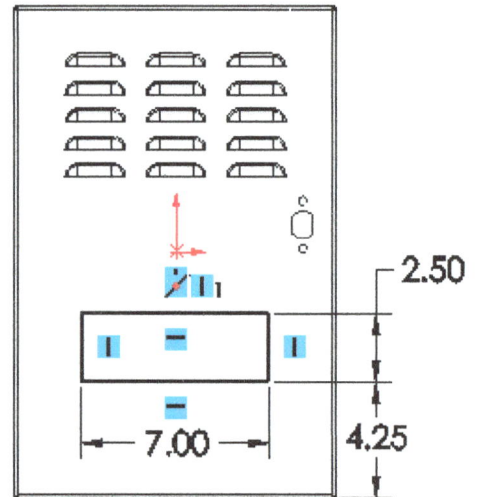

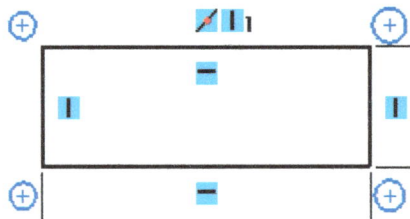

❖ Add a '7/32' dimension to the bottom left circle using the **Smart Dimension** icon in the CommandManager, or pull down the "Tools" menu and pick **Dimensions – Smart**.

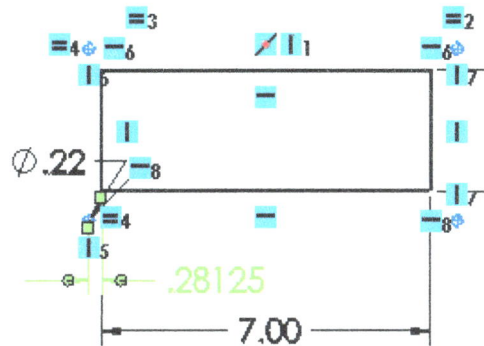

[=] Press the **Escape** key. Then, select the dimensioned circle. Hold down the **Ctrl** key and select the other three circles. In the **Properties** PropertyManager, click the **Equal** button.

[|] Select the centerpoint of the dimensioned 7/32 circle. Hold down the **Ctrl** key and select the centerpoint of the top left circle. In the **Properties** PropertyManager, click the **Vertical** button.

[—] Select the centerpoint of the top right circle. Hold down the **Ctrl** key and select the centerpoint of the top left circle. In the **Properties** PropertyManager, click the **Horizontal** button.

[|] Select the centerpoint of the bottom right circle. Hold down the **Ctrl** key and select the centerpoint of the top right circle. In the **Properties** PropertyManager, click the **Vertical** button.

[—] Select the centerpoint of the bottom left circle. Hold down the **Ctrl** key and select the centerpoint of the bottom right circle. In the **Properties** PropertyManager, click the **Horizontal** button.

❖ Add a '9/32' (.28125) horizontal dimension between the left side of the rectangle and the bottom left circle as shown using the **Smart Dimension** icon in the CommandManager, or pull down the "Tools" menu and pick **Dimensions – Smart**.

Double click on the .28125 dimension.

In the **Modify** dialog box, click on the down arrow and pick **Link Value** from the menu.

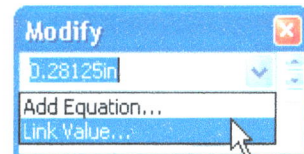

In the **Shared Values** dialog box, under **Name**, type 'MountHole'.

Click the **OK** button.

❖ Add a vertical dimension between the bottom line of the rectangle and the bottom left circle.

In the **Modify** dialog box, click on the down arrow and pick **Link Value** from the menu.

In the **Shared Values** dialog box, under **Name**, click on the down arrow. Pick **MountHole** from the list, and click the **OK** button.

✎ Add a vertical dimension between the top line of the rectangle and the top left circle. Link the value to **MountHole** again.

✎ Add a horizontal dimension between the right line of the rectangle and the bottom right circle. Link the value to **MountHole**.

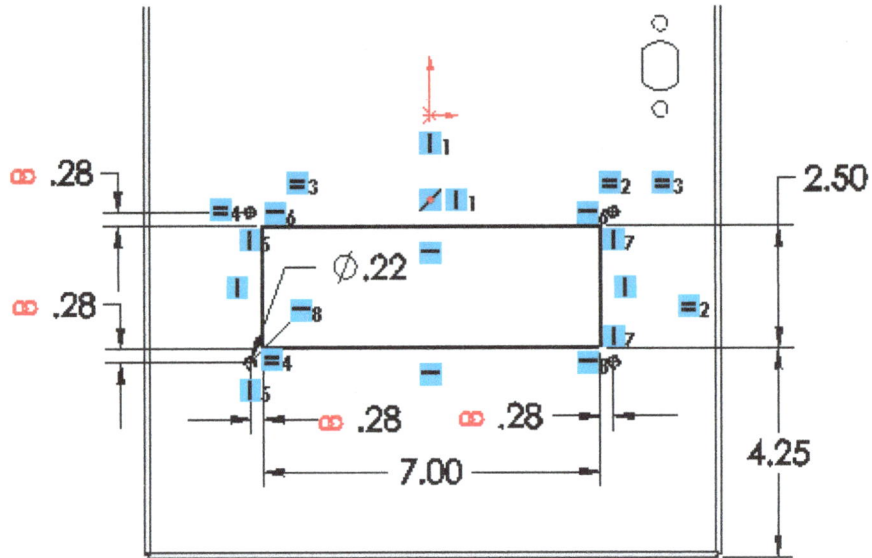

✎ Exit the sketch by clicking the **Exit Sketch** icon in the CommandManager or in the upper right corner of the graphics area.

✔ In the **Cut-Extrude** PropertyManager, check the **Link to thickness** check box and then click the green check mark button.

Create a Boundary Sketch

✎ Click the **Sketch** icon in the control area of the CommandManager. Then, click the **Sketch** icon or pull down the "Insert" menu and pick **Sketch**.

Select the front of the part for the plane to sketch on.

▭ Create a rectangle below the previous cut using the **Rectangle** icon in the CommandManager, or pull down the "Tools" menu and pick **Sketch Entities – Rectangle**.

Press the **Escape** key to deselect the command.

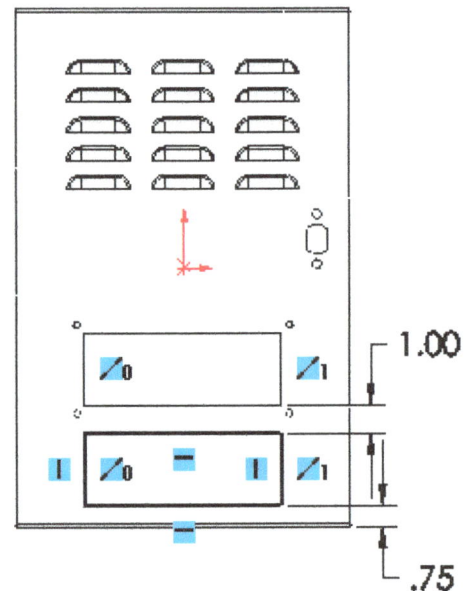

Select the left vertical line of the previous cut out. Then, **Ctrl** select the left vertical line of the sketched rectangle. In the **Properties** PropertyManager, click the **Collinear** button.

Select the right vertical line of the previous cut out. Then, **Ctrl** select the right vertical line of the sketched rectangle. In the **Properties** PropertyManager, click the **Collinear** button.

Add the '1' and '.75' vertical dimensions shown using the **Smart Dimension** icon in the CommandManager, or pull down the "Tools" menu and pick **Dimensions – Smart**.

Exit the sketch by clicking the **Exit Sketch** icon in the CommandManager or in the upper right corner of the graphics area.

Create a Fill Pattern

Pull down the "Insert" menu and pick **Pattern/Mirror – Fill Pattern**.

In the graphics area, select the rectangle sketch created in the previous step (Sketch13).

In the **Fill Pattern** PropertyManager, under **Pattern Layout**, make sure that the **Perforation** button is depressed.

Set the **Instance Spacing** to '.65'.

Click in the **Pattern Direction** box. In the graphics area, select the top half of the right vertical sketch line.

Under **Features to Pattern**, click on the **Create seed cut** radio button.

Click on the **Polygon** button.

Set the **Polygon Sides** to '6'.

Set the **Outer Radius** to '.25'.

In the **Fill Pattern** PropertyManager, click the green check mark button.

In the FeatureManager design tree, right click on rectangle sketch created in the previous step (**Sketch13**) and pick **Hide**.

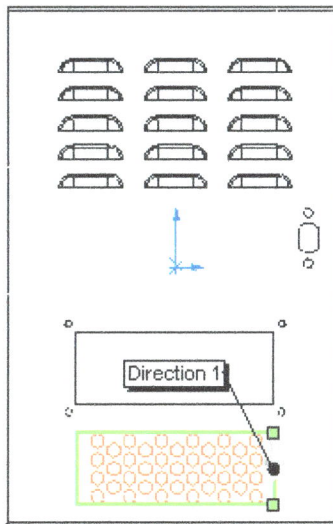

Create Hinge Mounting Holes

In the bottom left corner of your graphics area, change the View orientation by clicking the pull down arrow and picking **Bottom**, or press **Ctrl-6**.

In the graphics area, select the bottom of the part. A 2D sketch is created when you pre-select a planar face for the Hole Wizard. A 3D sketch is created if you first click Hole Wizard and then select a face.

Click the **Features** icon in the control area of the CommandManager. Then, click the **Hole Wizard** icon from the toolbar, or pull down the "Insert" menu and pick **Features – Hole – Wizard**.

Hole Wizard
Inserts a hole using a pre-defined cross-section.

In the **Hole Specification** PropertyManager, under the **Type** tab, click the **Hole** button.

From the **Standard** pull down list select '**Ansi Inch**'.

From the **Size** pull down list select '**15/64**'.

Under **End Condition**, pick **Up to Next** from the pull down list.

Next, click the **Positions** tab.

In the graphics area, click on the bottom of the part to place a hole to the right.

Add the dimensions shown below using the **Smart Dimension** icon in the CommandManager, or pull down the "Tools" menu and pick **Dimensions – Smart**.

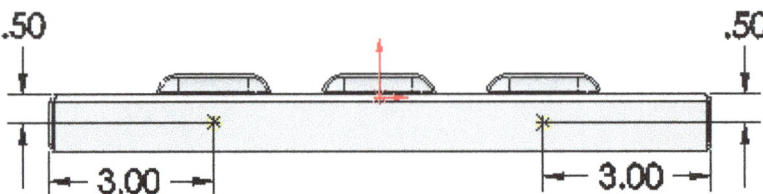

.50   .50

3.00   3.00

In the **Hole Specification** PropertyManager, click the green check mark button.

Mirror the Hinge Mounting Holes to the Top

Click the **Features** icon in the control area of the CommandManager. Then, select the **Mirror** icon from the toolbar, or pull down the "Insert" menu and select **Pattern/Mirror – Mirror**.

In the bottom left corner of your graphics area, change the View orientation by clicking the pull down arrow and picking **Trimetric**.

In the flyout FeatureManager design tree, select **Top Plane** as the **Mirror Face/Plane**.

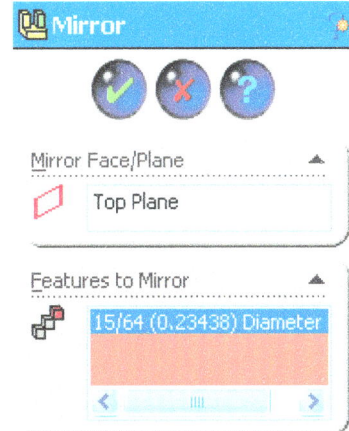

Under **Features to Mirror**, select **15/64 (0.23438) Diameter Hole1** from the flyout FeatureManager design tree.

A preview will appear as you select the features to ensure that you are selecting the correct features.

Click the green check mark button at the top of the **Mirror** PropertyManager to accept the settings and create the feature.

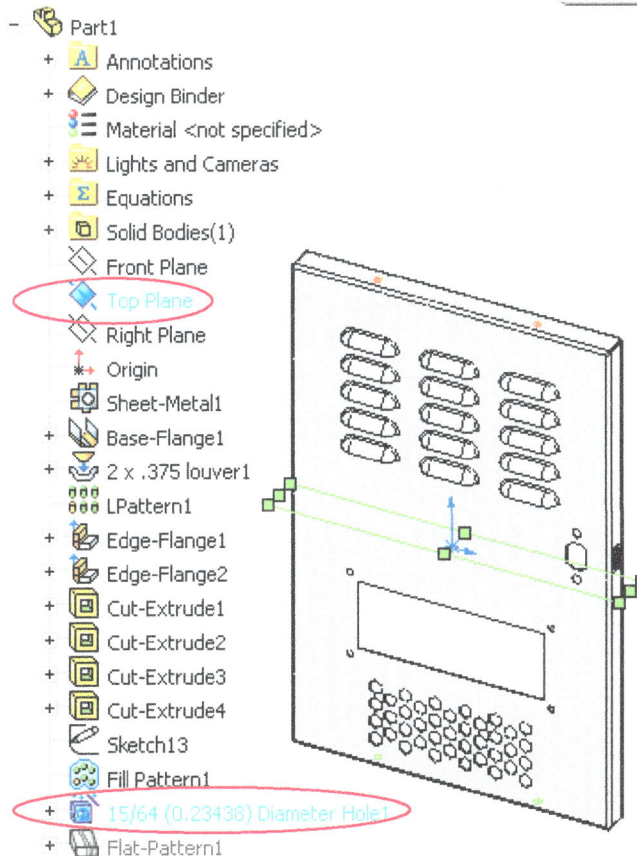

Saving the Part

🖫 Click the **Save** icon in the "Standard" toolbar, or pick **Save** from the "File" pull down menu.

In the **Save As** dialog box, in the **File name** box, type '**Locker Door**' and click **Save**.

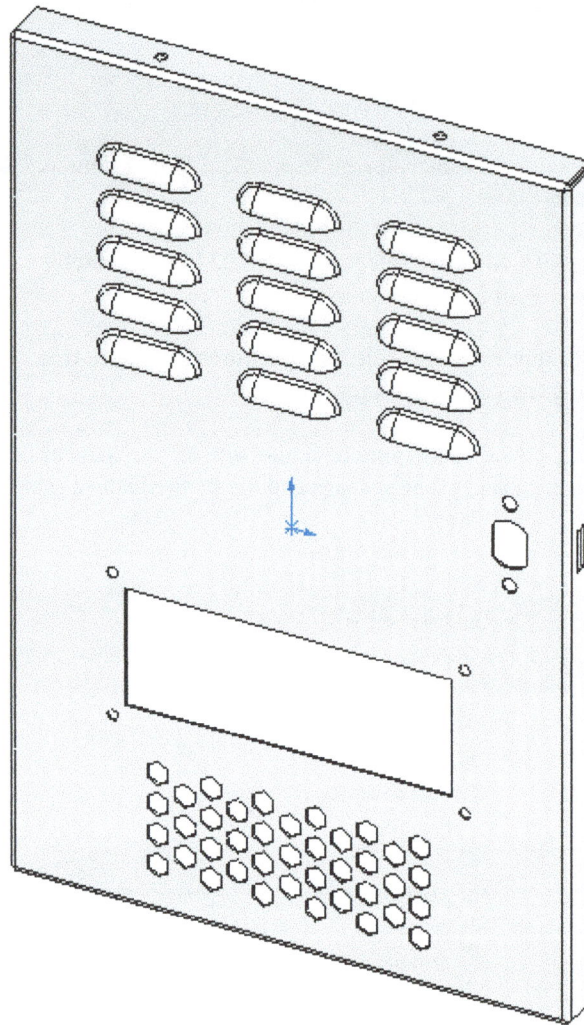

Chapter 9

Chassis

The Chassis shows you several new techniques, starting with the Vent command as a unique way to create cutouts for the cooling fans. What would have been a long process to sketch is greatly simplified by this feature.

The Linear Pattern function is then used to create the multiples, and the instances to skip feature lets you eliminate the occurrence you did not need.

This is another opportunity to use the Design Library. The RS-232, double lance, dimple and tombstone objects are all incorporated in this part. Again, patterns are used to place these objects and make future updates easier.

There should probably be another 20-30 round holes in this part, but you should already know how to create and place them.

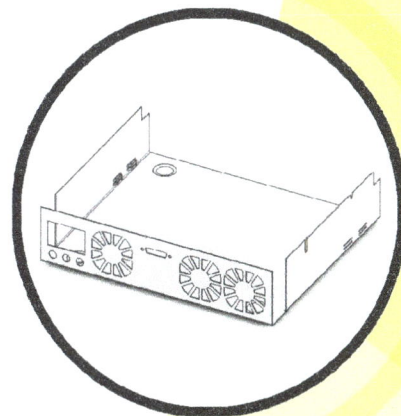

Create the Base Flange

Begin a new **Part** document by clicking the **New** icon in the "Standard" toolbar, or pull down the "File" menu and pick **New**.

Create a base flange by clicking the **Sheet Metal** icon in the control area of the CommandManager. Then, click the **Base-Flange/Tab** icon from the toolbar, or pull down the "Insert" menu and pick **Sheet Metal – Base Flange**.

Select the **Top** plane when prompted to select a plane on which to sketch the feature cross-section.

Create a rectangle with the origin inside the rectangle using the **Rectangle** icon in the CommandManager, or pull down the "Tools" menu and pick **Sketch Entities – Rectangle**.

Create a construction line diagonally across the rectangle by clicking the **Centerline** icon in the CommandManager, or pull down the "Tools" menu and pick **Sketch Entities – Centerline**.

Select the top left corner and then the bottom right corner of the rectangle that you just created to create the centerline as shown below.

Right click in the graphics area and pick **Select** from the menu, or press the **Escape** key.

Select the diagonal line. Then, hold down the **Ctrl** key and select the origin.

In the **Properties** PropertyManager, under **Add Relations**, click the **Midpoint** button.

Add a '**17**' horizontal dimension to the bottom line and a '**15**' vertical dimension to the left vertical line using the **Smart Dimension** icon in the CommandManager, or pull down the "Tools" menu and pick **Dimensions – Smart**.

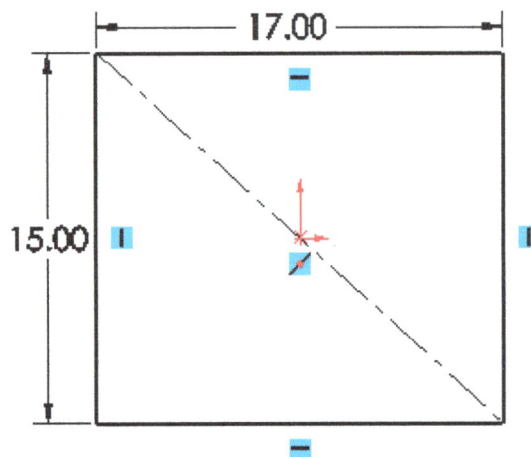

Exit the sketch by clicking the **Exit Sketch** icon in the CommandManager or in the upper right corner of the graphics area.

In the **Base Flange** PropertyManager under **Sheet Metal Parameters**, set the **Thickness** to '.**0598**'. Make sure that the **Reverse direction** check box is checked.

Click the green check mark button at the top of the **Base Flange** PropertyManager to accept the settings and create the part.

Add an Edge Flange

Click the **Edge Flange** icon in the CommandManager, or pull down the "Insert" menu and pick **Sheet Metal – Edge Flange**.

Select the front of the part and move the cursor up and click to set the direction of the flange.

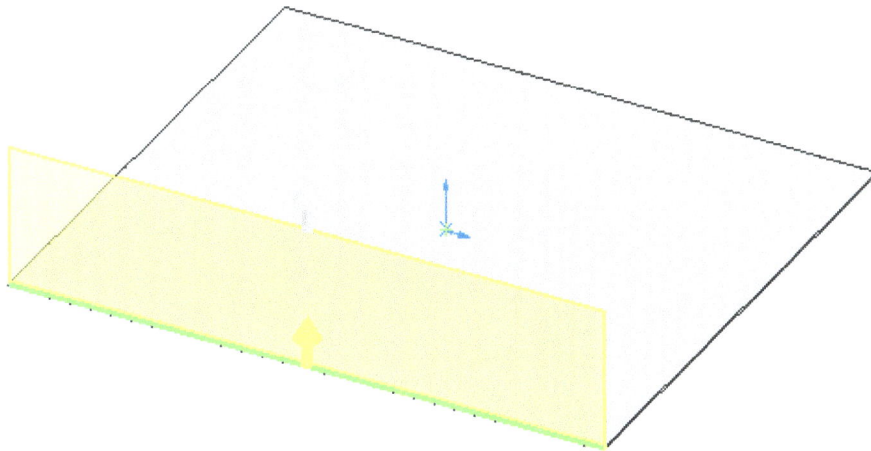

In the **Edge-Flange** PropertyManager, set the **Flange Length** to **Blind** and the **Length** to '**4**'.

Click the **Outer Virtual Sharp** button and the **Material Inside Flange Position** button.

Click the green check mark button at the top of the **Edge-Flange** PropertyManager to accept the settings and create the flange.

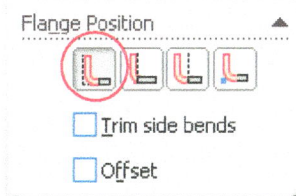

Add Another Edge Flange

Click the **Edge Flange** icon in the CommandManager again, or pull down the "Insert" menu and pick **Sheet Metal – Edge Flange**.

Select the right of the part and move the cursor up and click to set the direction of the flange.

In the **Edge-Flange** PropertyManager, set the **Flange Length** to **Blind** and the **Length** to '**4**'.

Click the **Outer Virtual Sharp** button and the **Material Inside Flange Position** button.

In the **Edge-Flange** PropertyManager, click the **Edit Flange Profile** button. Edit Flange Profile

Left click and drag the left vertical side line in towards the middle.

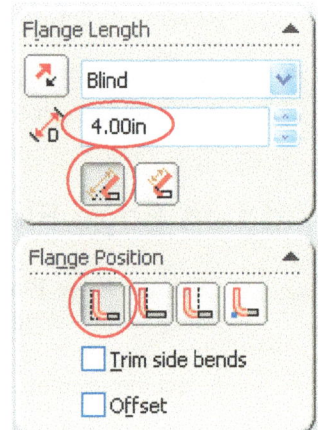

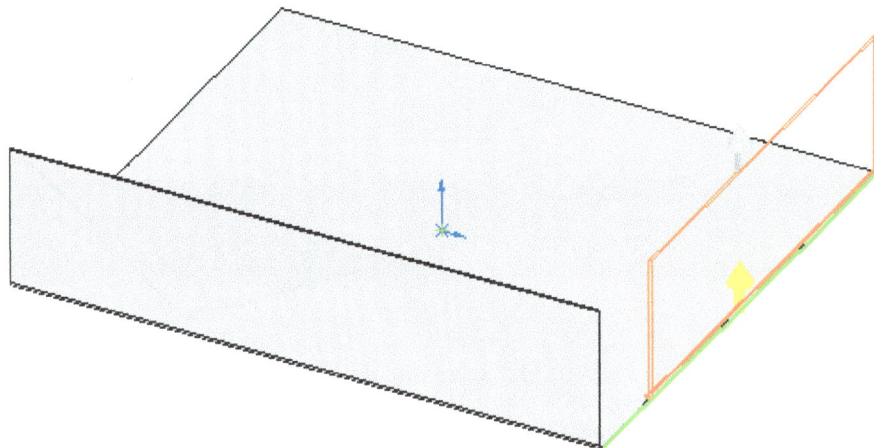

In the bottom left corner of the graphics area, change the View orientation by clicking the pull down arrow and picking **Right**, or press **Ctrl-4**. *Right

Then, create a notch in the upper right corner of the sketch as shown using the **Line** icon in the CommandManager, or pull down the "Tools" menu and pick **Sketch Entities – Line**.

Next, trim the notch using the **Trim Entities** icon or pull down the "Tools" menu and pick **Sketch Tools - Trim Entities**.

Add a '**2**' horizontal dimension to the left line and two '**1**' dimensions to the notch using the **Smart Dimension** icon in the CommandManager, or pull down the "Tools" menu and pick **Dimensions – Smart**.

In the **Profile Sketch** dialog box, click the **Finish** button.

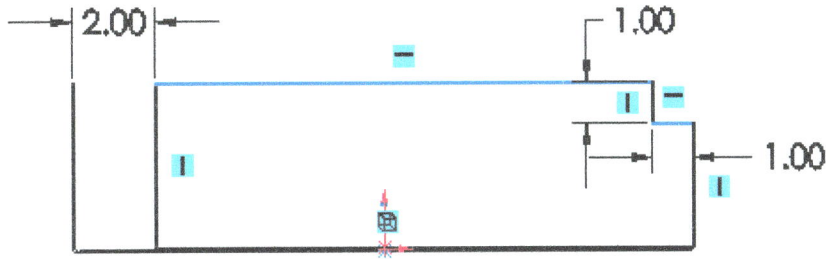

Mirror the Flange

Click the **Features** icon in the control area of the CommandManager. Then, select the **Mirror** icon from the toolbar, or pull down the "Insert" menu and select **Pattern/Mirror – Mirror**.

In the bottom left corner of the graphics area, change the View orientation by clicking the pull down arrow and picking **Trimetric**.

In the flyout FeatureManager design tree, select **Right Plane** as the **Mirror Face/Plane**.

Under **Features to Mirror**, select **Edge-Flange2** from the flyout FeatureManager design tree.

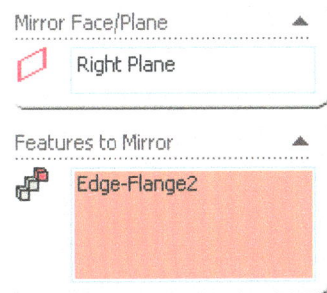

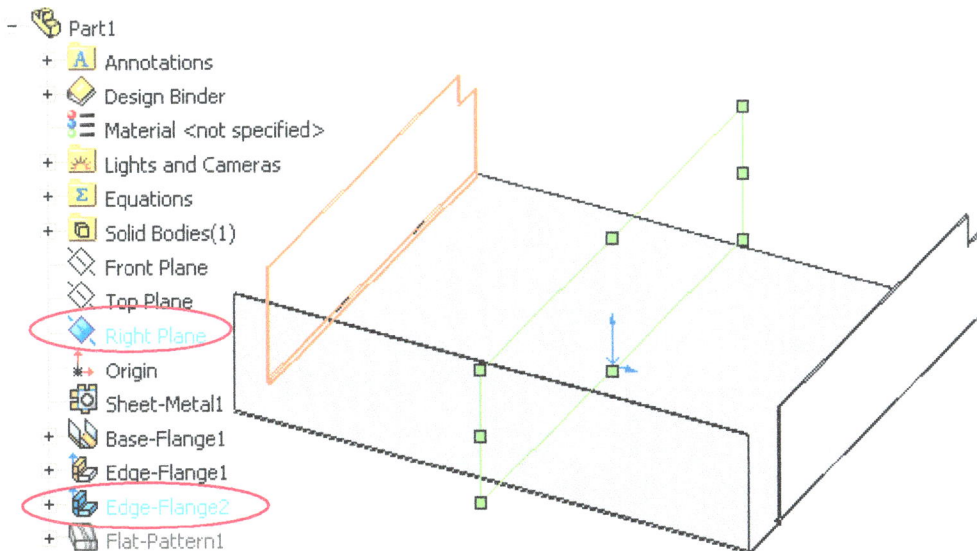

Click the green check mark button at the top of the **Mirror** PropertyManager to accept the settings and create the feature.

Create a Vent Sketch

Click the **Sketch** icon in the control area of the CommandManager. Then, click the **Sketch** icon or pull down the "Insert" menu and pick **Sketch**.

Select the front of the part as the sketching plane. In the bottom left corner of the graphics area, change the View orientation by clicking the pull down arrow and picking **Front**,

Create a '3' by '3' rectangle '.5' from the upper right hand corner of the front of the part using the **Rectangle** icon in the CommandManager, or pull down the "Tools" menu and pick **Sketch Entities – Rectangle**.

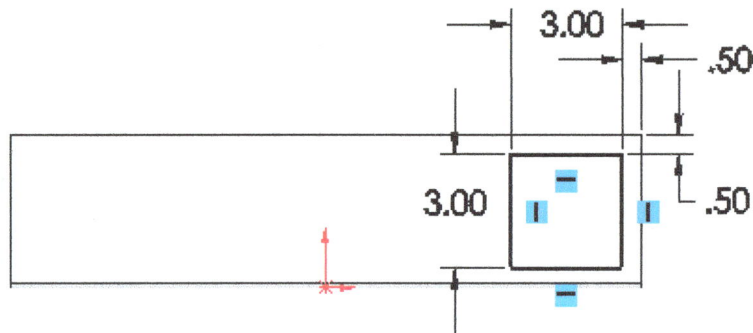

Click the **Fillet** icon in the CommandManager, or pull down the "Tools" menu and pick **Sketch Tools – Fillet**.

In the **Sketch Fillet** PropertyManager, enter a **Radius** of '1' and select the four corners of the rectangle.

Create a vertical centerline line through the center of the rectangle connecting the midpoint of the top and bottom horizontal lines of the rectangle using the **Centerline** icon in the CommandManager, or pull down the "Tools" menu and pick **Sketch Entities – Centerline**.

Create a horizontal line starting at the midpoint of the centerline and ending at the midpoint of the left vertical line of the rectangle using the **Line** icon in the CommandManager, or pull down the "Tools" menu and pick **Sketch Entities – Line**.

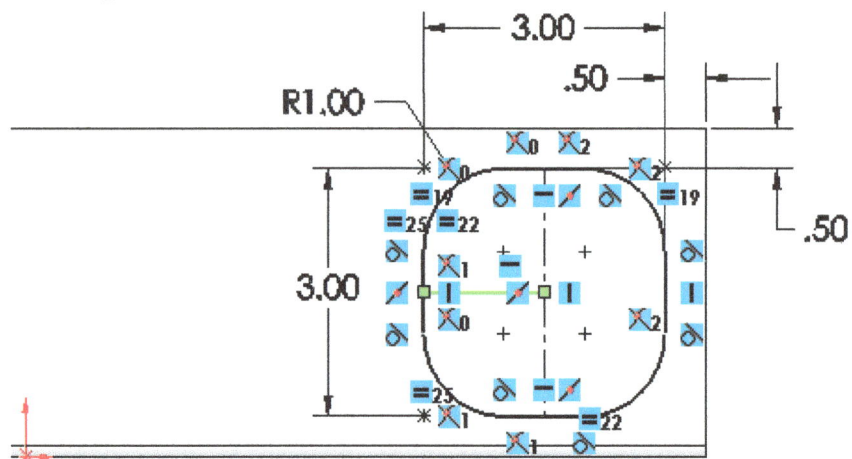

Pull down the "Tools" menu and select **Sketch Tools – Circular Pattern**.

In the **Circular Pattern** PropertyManager, under **Entities to Pattern**, the horizontal sketch line should be selected.

Circular sketch patterns are centered to the sketch origin by default. To change this, under **Parameters**, click in the first box. Then, in the graphics area, select the midpoint of the centerline.

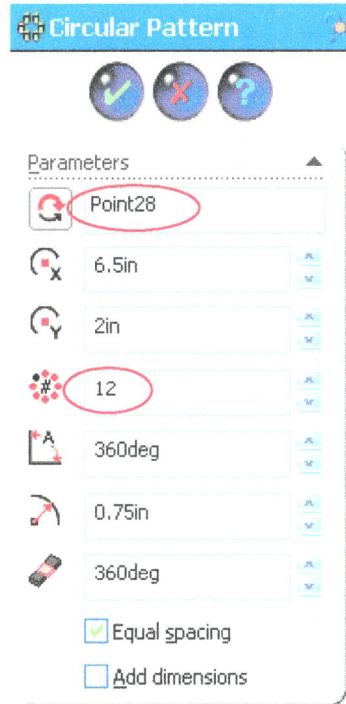

Set the **Number** of pattern instances to '**12**'.

Make sure the **Spacing** is set to '**360**' and that the **Equal spacing** check box is checked.

Click the green check mark button at the top of the **Circular Pattern** PropertyManager.

To see the lines better, pull down the "View" menu and pick **Sketch Relations** to hide the sketch relations. You may also want to zoom in using **Zoom to Area**.

Next, extend the angled lines using the **Trim Entities** icon or pull down the "Tools" menu and pick **Sketch Tools - Trim Entities**.

In the **Trim** PropertyManager, click on the **Power trim** button.

Hold down the **Shift** key. Then, press the mouse button and drag the cursor over the outer ends of the two angled lines. Make sure that you do not touch the vertical or horizontal lines. Do this for the four sets of angled lines as shown.

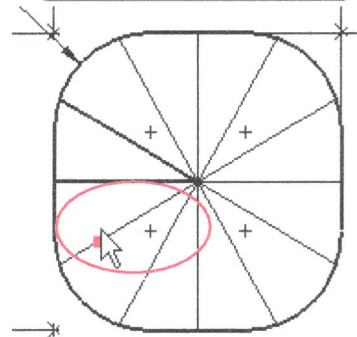

Create a circle in the middle of the sketch using the **Circle** icon in the CommandManager, or pull down the "Tools" menu and pick **Sketch Entities – Circle**.

Add the '**1.5**' dimater dimension to the circle using the **Smart Dimension** icon in the CommandManager, or pull down the "Tools" menu and pick **Dimensions – Smart**.

Exit the sketch by clicking the **Exit Sketch** icon in the CommandManager or in the upper right corner of the graphics area.

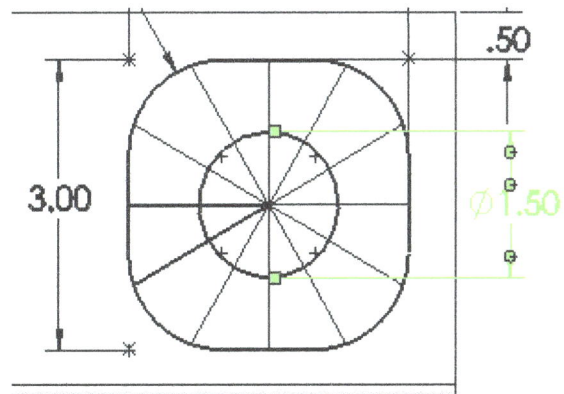

Insert the Fastening Feature

Pull down the "Insert" menu and pick **Fastening Feature – Vent**.

In the graphics area, select the outside profile for the **Boundary** (the original rectangle and fillets). You must select all eight entities.

In the **Vent** PropertyManager, click in the **Ribs** box. Then, in the graphics area, select the 12 lines that were created by the circular pattern.

Set the **width of the ribs** to '**.2**'.

Click in the **Fill-In Boundary** box. Then, in the graphics area, select the circle.

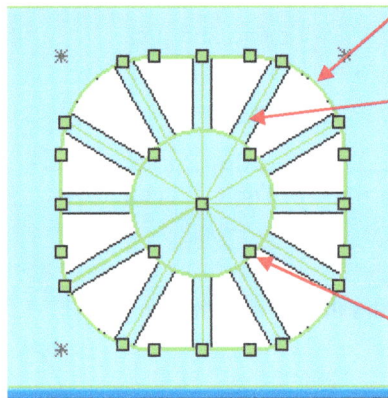

Click the green check mark button at the top of the **Vent** PropertyManager to accept the settings and create the feature.

Create a Linear Pattern

Click the **Features** icon in the control area of the CommandManager. Then, click the **Linear Pattern** icon from the CommandManager, or pull down the "Insert" menu and pick **Pattern/Mirror – Linear Pattern**.

In the graphics area, select the right side of the top edge of the part to set the **Direction 1** of the pattern as shown.

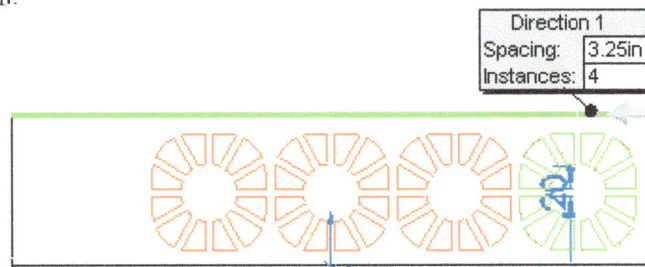

Direction 1	
Spacing:	3.25in
Instances:	4

SolidWorks for the Sheet Metal Guy

In the **Linear Pattern** PropertyManager, set the **Spacing** to '**3.25**' and the **Number** to '**4**'.

Click in the **Features to Pattern** box. In the flyout FeatureManager design tree, select **Vent1**.

In the **Linear Pattern** PropertyManager, click in the **Instances to Skip** box.

In the graphics area, click on the center point of the third from the right vent as shown for the instance to skip.

Click the green check mark button at the top of the **Linear Pattern** PropertyManager.

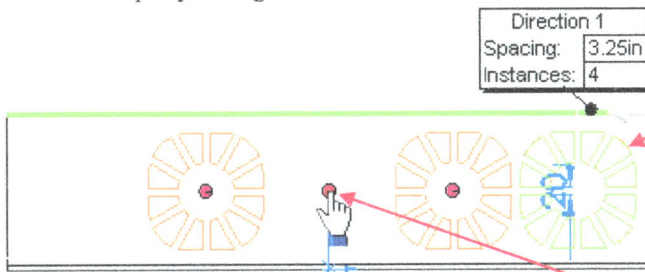

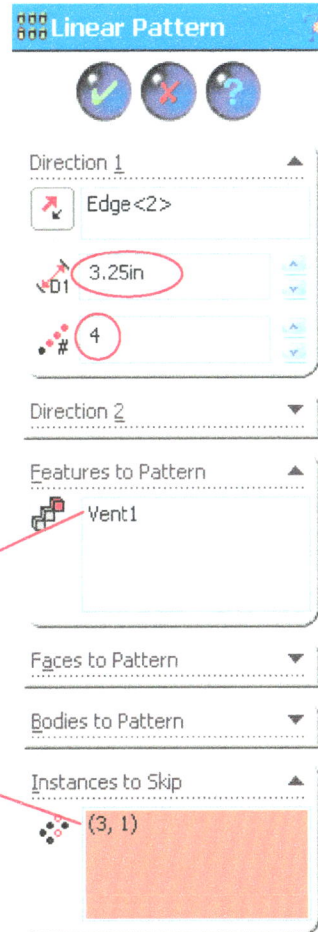

Insert a Punching Shape

Click the **Design Library** icon in the Task Pane to expand the **Design Library** tab.

In the **SheetMetalGuy** Design Library, click on your **Punching Shapes** folder to display the folder contents in the lower pane of the **Design Library** tab.

Drag **RS-232** from the lower pane of the **Design Library** tab onto the front of the part as shown.

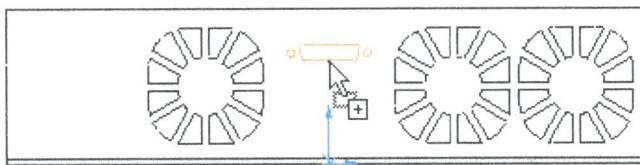

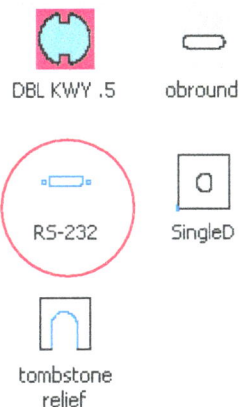

In the **RS-232** PropertyManager, under **Location**, click the **Edit Sketch** button.

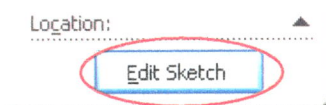

Click the **Smart Dimension** icon in the CommandManager, or pull down the "Tools" menu and pick **Dimensions – Smart**.

Add a '**8.5**' horizontal dimension from the right edge to the centerline of the RS-232. Then, add a '**1**' vertical dimension from the top edge line to the center of the left circle as shown. You may need to zoom in to select the appropriate entities.

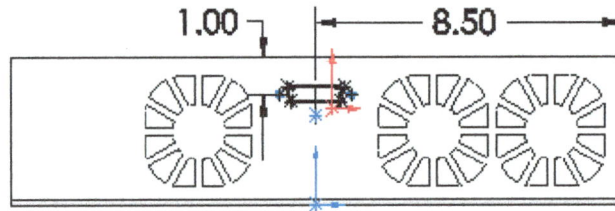

In the **Library Feature Profile** dialog box, click the **Finish** button.

Create an Extruded Cut

Click the **Sheet Metal** icon in the control area of the CommandManager. Then, click the **Extruded Cut** icon or pull down the "Insert" menu and pick **Cut – Extrude**.

Select the front of the part as shown.

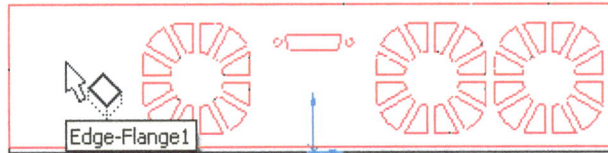

Create a '**1.75**' tall by '**2.75**' wide rectangle on the left side of the front of the part using the **Rectangle** icon in the CommandManager, or pull down the "Tools" menu and pick **Sketch Entities – Rectangle**.

Create three circles below the rectangle using the **Circle** icon in the CommandManager, or pull down the "Tools" menu and pick **Sketch Entities – Circle**.

Press the **Escape** key and then select the centerpoint of the left circle. Hold down the **Ctrl** key and select the centerpoint of the other circles. In the **Properties** PropertyManager, click on the **Horizontal** button.

Select the left circle, and then, hold down the **Ctrl** key and select the other circles. In the **Properties** PropertyManager, click on the **Equal** button.

Right click on the bottom horizontal line of the rectangle and pick **Select Midpoint**. Hold down the **Ctrl** key and select the centerpoint of the circle in the middle. In the **Properties** PropertyManager, click on the **Vertical** button.

Add the rest of the dimensions using the **Smart Dimension** icon in the CommandManager, or pull down the "Tools" menu and pick **Dimensions – Smart**.

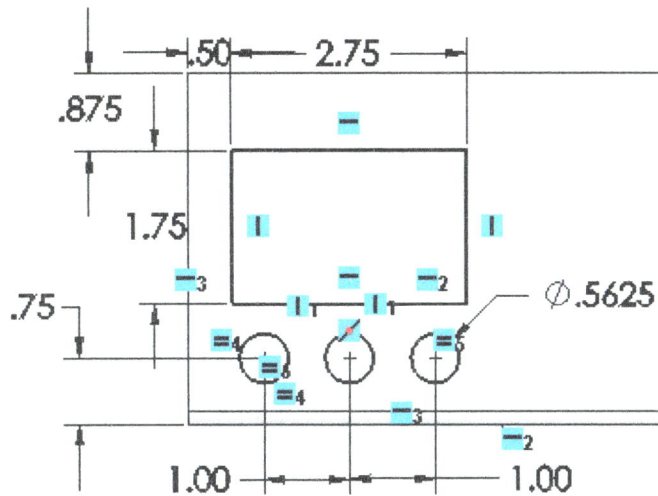

Exit the sketch by clicking the **Exit Sketch** icon in the CommandManager or in the upper right corner of the graphics area.

In the **Cut-Extrude** PropertyManager, check the **Link to thickness** check box and then click the green check mark button at the top of the **Cut-Extrude** PropertyManager.

Insert Your Double Lance

In the bottom left corner of the graphics area, change the View orientation by clicking the pull down arrow and picking **Right**.

Click the **Design Library** icon in the Task Pane to expand the **Design Library** tab.

In the **SheetMetalGuy** Design Library, click on your **Forming Tools** folder to display the folder contents in the lower pane of the **Design Library** tab.

Drag **.125 wide double lance** from the lower pane of the **Design Library** tab onto the front of the part as shown.

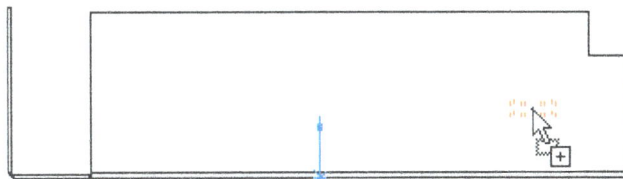

✎ Add '2.5' horizontal dimension and a '.5' vertical dimension as shown using the **Smart Dimension** icon in the CommandManager, or pull down the "Tools" menu and pick **Dimensions – Smart**.

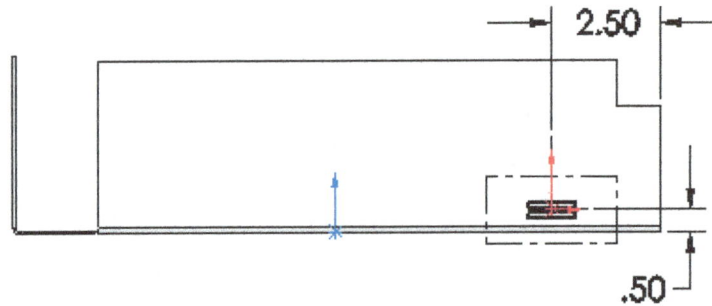

In the **Position form feature** dialog box, click the **Finish** button.

Create a Linear Pattern of the Double Lance

⣿ Click the **Features** icon in the control area of the CommandManager. Then, click the **Linear Pattern** icon from the CommandManager, or pull down the "Insert" menu and pick **Pattern/Mirror – Linear Pattern**.

In the graphics area, select the right side of the bottom edge of the part to set the **Direction 1** of the pattern as shown.

➹ In the **Linear Pattern** PropertyManager, set the **Spacing** to '2' and the **Number** to '2'.

Click in the **Features to Pattern** box. In the flyout FeatureManager design tree, select **.125 wide double lance1**.

✔ Click the green check mark button at the top of the **Linear Pattern** PropertyManager.

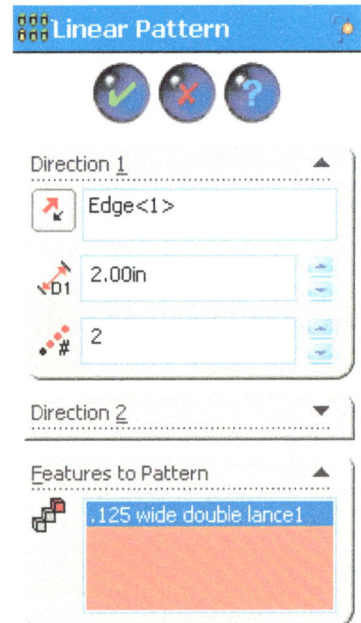

Insert a Dimple

In the bottom left corner of the graphics area, change the View orientation by clicking the pull down arrow and picking **Top**.

📖 Click the **Design Library** icon in the Task Pane to expand the **Design Library** tab.

In the **SheetMetalGuy** Design Library, click on your **Forming Tools** folder to display the folder contents in the lower pane of the **Design Library** tab.

Drag **dimple** from the lower pane of the **Design Library** tab onto the top of the part.

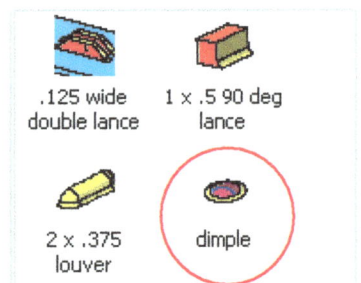

.125 wide double lance   1 x .5 90 deg lance

2 x .375 louver   dimple

Add '2' horizontal dimension and a '1.5' vertical dimension as shown using the **Smart Dimension** icon in the CommandManager, or pull down the "Tools" menu and pick **Dimensions – Smart**.

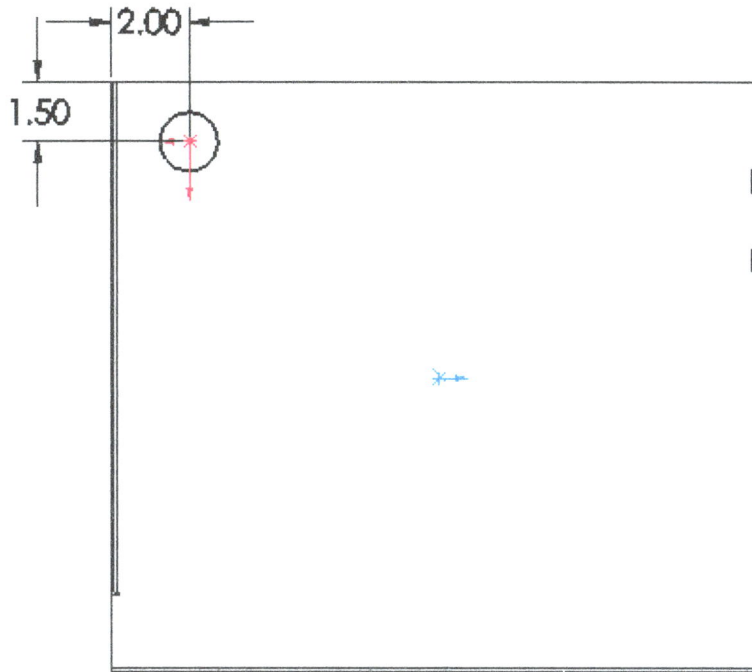

In the **Position form feature** dialog box, click the **Finish** button.

Create a Linear Pattern of the Dimple

Click the **Linear Pattern** icon from the CommandManager, or pull down the "Insert" menu and pick **Pattern/Mirror – Linear Pattern**.

In the graphics area, select the left side of the top edge of the part to set the **Direction 1** of the pattern as shown. In the **Linear Pattern** PropertyManager, set the **Spacing** to '13' and the **Number** to '2'.

Under **Direction 2**, click on the **Pattern Direction** box. Then, in the graphics area, select the top of the left edge of the part. In the **Linear Pattern** PropertyManager, set the **Spacing** to '12' and the **Number** to '2'.

Click in the **Features to Pattern** box. In the flyout FeatureManager design tree, select **dimple1**.

Click the green check mark button at the top of the **Linear Pattern** PropertyManager.

Linear Pattern

Direction 1
Edge<1>
D1   13.00in
2

Direction 2
Edge<3>
D2   12.00in
2
☐ Pattern seed only

Features to Pattern
dimple1

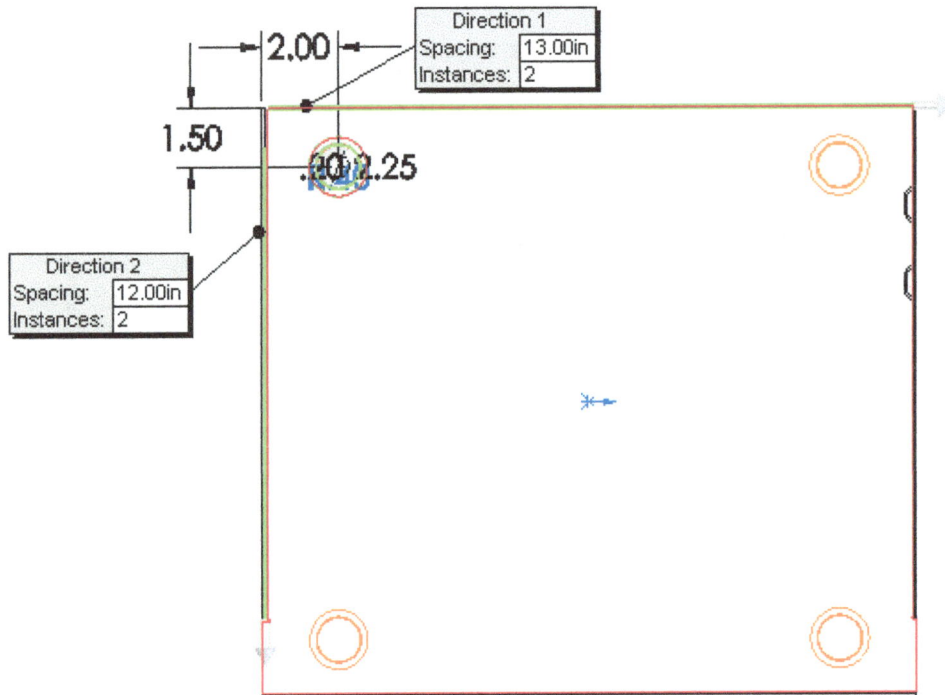

Direction 1
Spacing: 13.00in
Instances: 2

2.00

1.50

.125

Direction 2
Spacing: 12.00in
Instances: 2

Add Linear Pattern to the Mirror

In the FeatureManager design tree, left click and drag **.125 wide double lance1** below **Edge-Flange2** as shown.

Then, left click and drag **LPattern2** below **.125 wide double lance1** as shown.

Finally, right click on **Mirror1** and pick **Edit Feature**.

In the **Mirror1** PropertyManager, click in the **Features to Mirror** box.

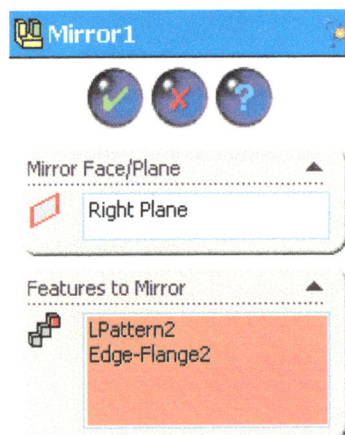

Mirror1

Mirror Face/Plane
Right Plane

Features to Mirror
LPattern2
Edge-Flange2

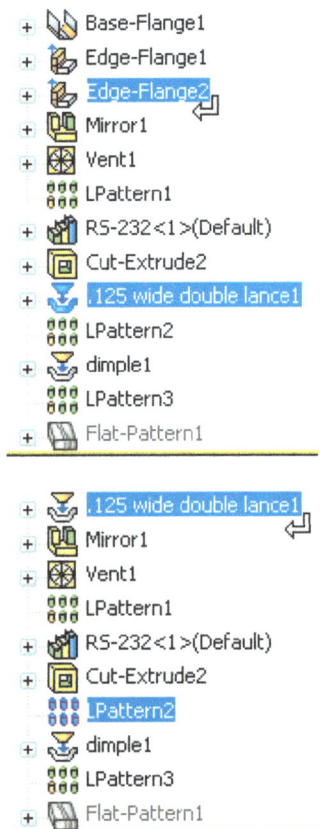

+ Base-Flange1
+ Edge-Flange1
+ Edge-Flange2
+ Mirror1
+ Vent1
  LPattern1
+ RS-232<1>(Default)
+ Cut-Extrude2
+ .125 wide double lance1
  LPattern2
+ dimple1
  LPattern3
+ Flat-Pattern1

+ .125 wide double lance1
+ Mirror1
+ Vent1
  LPattern1
+ RS-232<1>(Default)
+ Cut-Extrude2
  LPattern2
+ dimple1
  LPattern3
+ Flat-Pattern1

In the bottom left corner of the graphics area, change the View orientation by clicking the pull down arrow and picking **Trimetric**.

In the flyout FeatureManager design tree, select **LPattern2**.

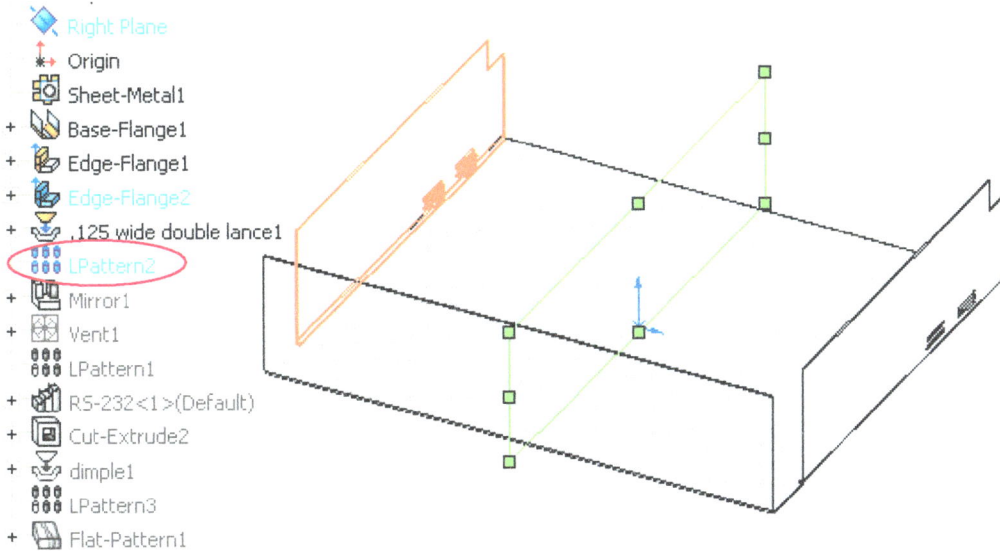

Click the green check mark button at the top of the **Mirror1** PropertyManager to accept the new settings.

Add Tombstone Reliefs

Click the **Design Library** icon in the Task Pane to expand the **Design Library** tab.

In the **SheetMetalGuy** Design Library, click on your **Punching Shapes** folder to display the folder contents in the lower pane of the **Design Library** tab.

Drag **tombstone relief** from the lower pane of the **Design Library** tab onto the right side of the part as shown.

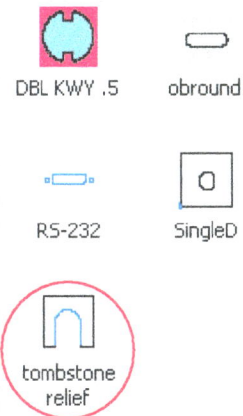

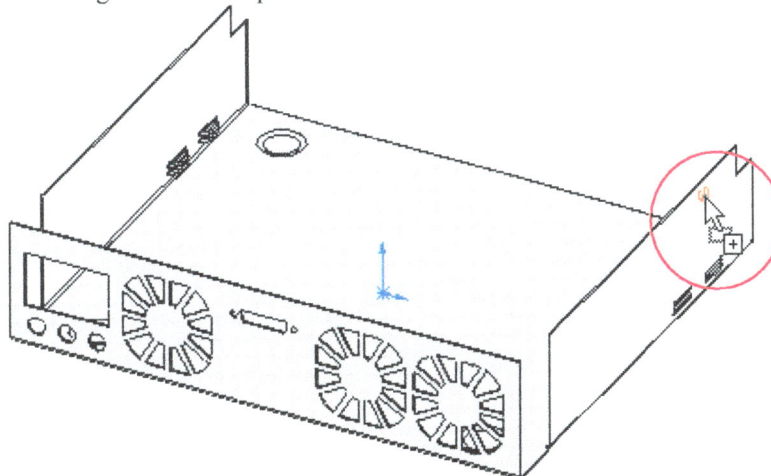

In the graphics area, select the top edge of the right flange as shown.

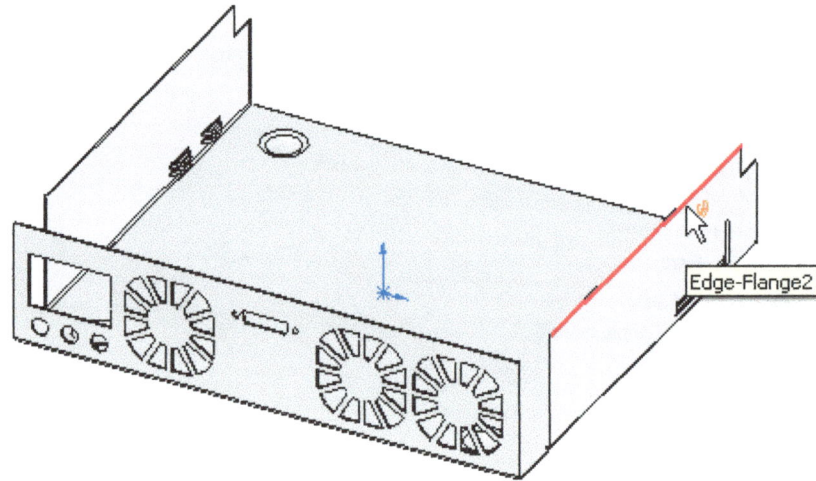

Edge-Flange2

Click the green check mark button at the top of the **tombstone relief** PropertyManager.

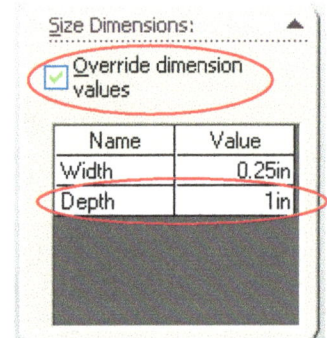

Drag another **tombstone relief** from the lower pane of the **Design Library** tab onto the right side of the part as shown.

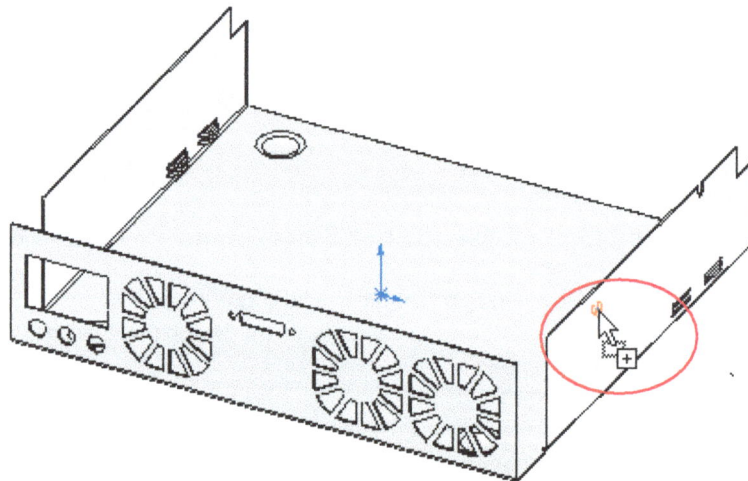

Again, in the graphics area, select the top edge of the right flange.

In the **tombstone relief** PropertyManager, under **Size Dimensions**, check the **Override dimension values** check box.

Change the **Depth** to '1'.

Click the green check mark button at the top of the **tombstone relief** PropertyManager.

Size Dimensions:

Override dimension values

Name	Value
Width	0.25in
Depth	1in

SolidWorks for the Sheet Metal Guy

In the FeatureManager design tree, click on the plus sign next to **tombstone relief <1>**.

Right click on the **Cut-Extrude1** and pick **Edit Sketch**.

In the bottom left corner of the graphics area, change the View orientation by clicking the pull down arrow and picking **Right**, or press **Ctrl-4**.

Add a '**4**' horizontal dimension from the right edge of the part to the centerline of the relief using the **Smart Dimension** icon in the CommandManager, or pull down the "Tools" menu and pick **Dimensions – Smart**.

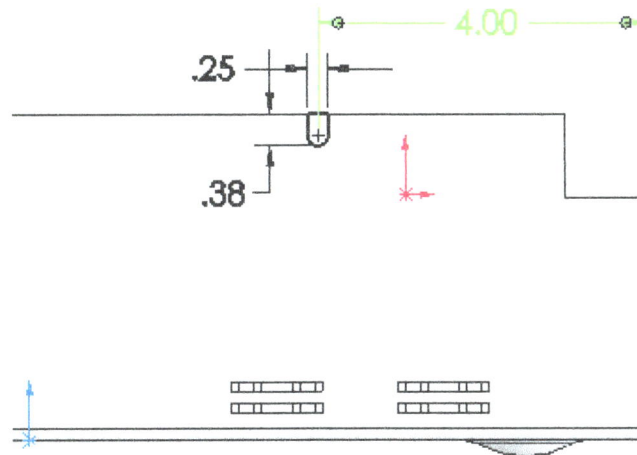

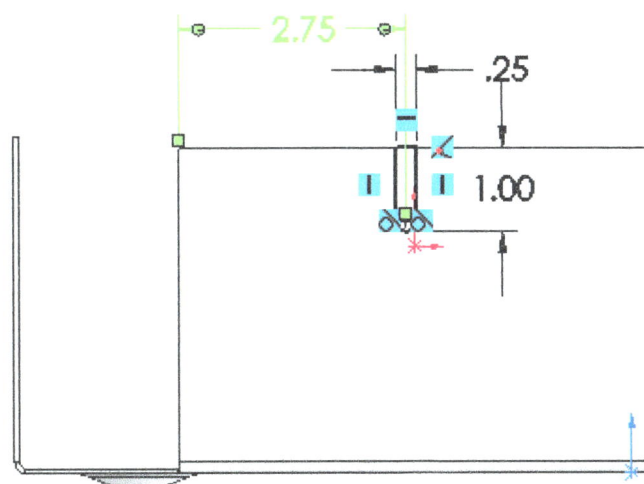

Exit the sketch by clicking the **Exit Sketch** icon in the CommandManager or in the upper right corner of the graphics area.

In the FeatureManager design tree, click on the plus sign next to **tombstone relief <2>**.

Right click on the **Cut-Extrude11** and pick **Edit Sketch**.

Add a '**2.75**' horizontal dimension from the left edge of the part to the centerline of the relief using the **Smart Dimension** icon in the CommandManager, or pull down the "Tools" menu and pick **Dimensions – Smart**.

Exit the sketch by clicking the **Exit Sketch** icon in the CommandManager or in the upper right corner of the graphics area.

In the bottom left corner of the graphics area, change the View orientation by clicking the pull down arrow and picking **Trimetric**.

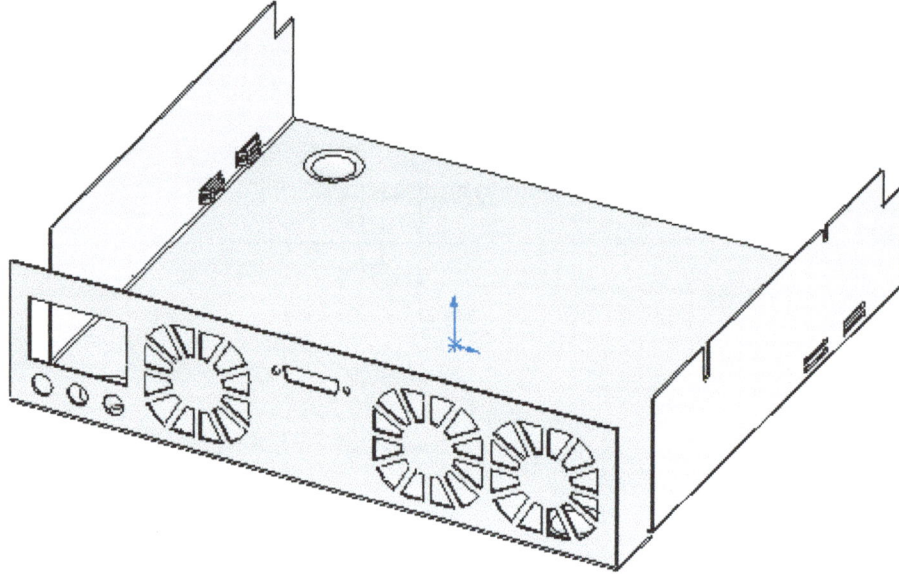

Saving the Part

Click the **Save** icon in the "Standard" toolbar, or pick **Save** from the "File" pull down menu.

In the **Save As** dialog box, in the **File name** box, type '**Chassis**' and click **Save**.

Chapter 10

Front Panel

One more time. This part provides a little of everything. But that is what it is all about. A final exam you might say. This part was originally created as a CNC program for a turret press manufacturer to use at a trade show. A lot of holes to keep the machine punching and some interesting patterns to get the attention of those watching. The keyboard hole pattern was left out to simplify the part a little.

When you create the arc pattern of the keyhole, pay attention to the shape rotating around the arc and the radio button for the **Alignment method** to hold the shape in its original orientation.

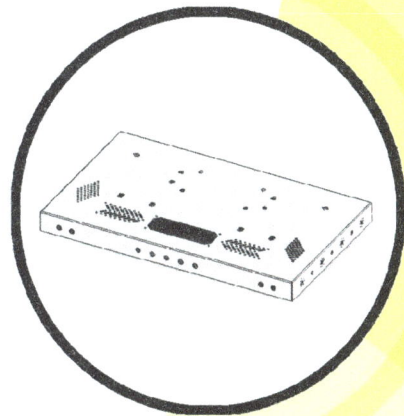

Create the Base Flange

Begin a new **Part** document by clicking the **New** icon in the "Standard" toolbar, or pull down the "File" menu and pick **New**.

Create a base flange by clicking the **Sheet Metal** icon in the control area of the CommandManager. Then, click the **Base-Flange/Tab** icon from the toolbar, or pull down the "Insert" menu and pick **Sheet Metal – Base Flange**.

Select the **Top** plane when prompted to select a plane on which to sketch the feature cross-section.

Create a rectangle with the origin inside the rectangle using the **Rectangle** icon in the CommandManager, or pull down the "Tools" menu and pick **Sketch Entities – Rectangle**.

Create a construction line diagonally across the rectangle by clicking the **Centerline** icon in the CommandManager, or pull down the "Tools" menu and pick **Sketch Entities – Centerline**.

Select the top left corner and then the bottom right corner to create the centerline.

Press the **Escape** key to deselect the **Centerline** tool.

Select the diagonal line. Then, hold down the **Ctrl** key and select the origin.

In the **Properties** PropertyManager, under **Add Relations**, click the **Midpoint** button.

Click the **Smart Dimension** icon in the CommandManager, or pull down the "Tools" menu and pick **Dimensions – Smart**.

Add a '**24**' vertical dimension to the left vertical line and a '**40**' horizontal dimension between the left and right side of the sketch.

Press the **f** key on the keyboard to **Zoom to Fit** so you can see the entire rectangle and center it in the graphics area. (Keyboard shortcut key, **Zoom to Fit**: f)

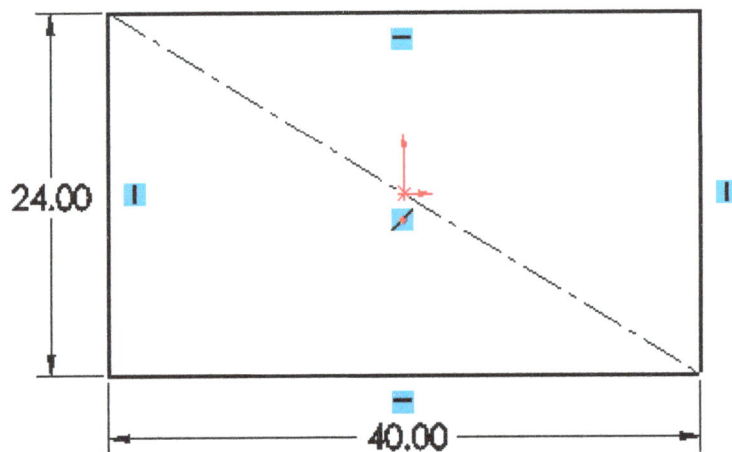

Exit the sketch by clicking the **Exit Sketch** icon in the CommandManager or in the upper right corner of the graphics area.

In the **Base Flange** PropertyManager, under **Sheet Metal Parameters**, set the **Thickness** to '**.0747**'. Make sure that the **Reverse direction** check box is <u>not</u> checked.

Click the green check mark button at the top of the **Base Flange** PropertyManager to accept the settings and create the part.

Create the Edge Flanges

Click the **Edge Flange** icon in the CommandManager, or pull down the "Insert" menu and pick **Sheet Metal – Edge Flange**.

Select the front left edge of the base flange and move the cursor down and click to set the direction of the flange.

Then, select the other three edges as shown.

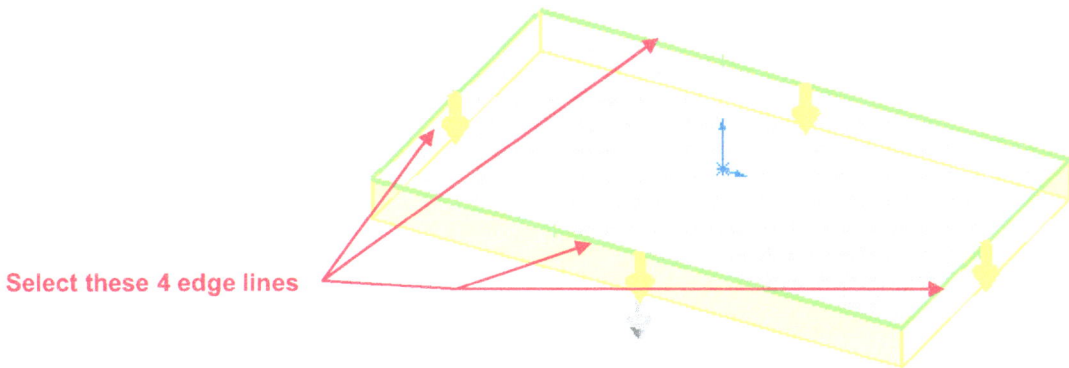

Select these 4 edge lines

In the **Edge-Flange** PropertyManager, set the **Flange Length** to **Blind** and the **Length** to '**3**'.

Click the **Outer Virtual Sharp** button and the **Material Inside Flange Position** button.

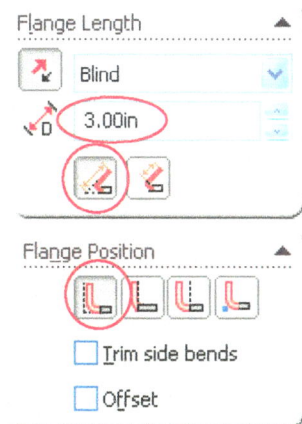

Click the green check mark button at the top of the **Edge-Flange** PropertyManager to accept the settings and create the flange.

Create a Line of Holes on the End Flanges

Select the right end flange for the plane to sketch on, by picking on the flange near the left end of the flange as shown. This is where the first hole will be located. Also, remember selecting the flange before starting the command causes the Hole Wizard to use a 2D sketch instead of a 3D sketch and will ensure that your dimensions are correct.

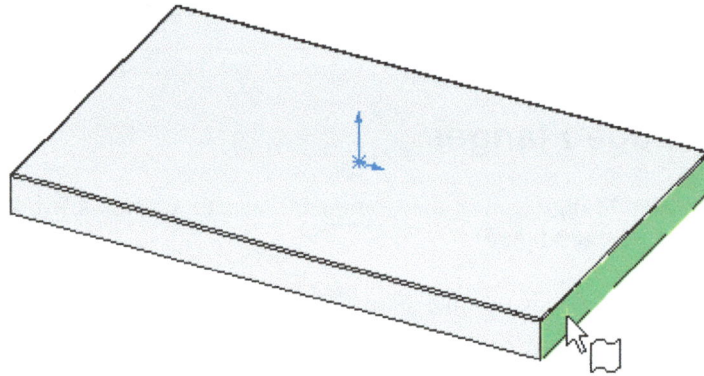

Click the **Features** icon in the control area of the CommandManager. Then, click the **Hole Wizard** icon in the CommandManager, or pull down the "Insert" menu and pick **Features – Hole – Wizard**.

In the **Hole Specification** PropertyManager, under the **Type** tab, click the **Hole** button.

From the **Size** pull down list select **5/8**.

Under **End Condition**, pick **Up to Next** from the pull down list.

Click the green check mark button to accept the settings and create the hole.

Click the **Linear Pattern** icon in the CommandManager, or pull down the "Insert" menu and pick **Pattern/Mirror – Linear Pattern**.

In the bottom left corner of the graphics area, change the View orientation by clicking the pull down arrow and picking **Right**, or press **Ctrl-4**.

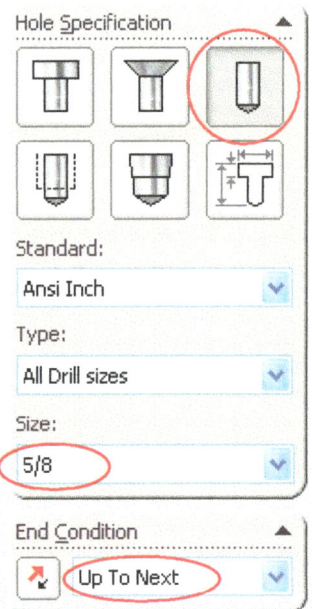

*Right

Hole Specification

Hole Specification

Standard:
Ansi Inch

Type:
All Drill sizes

Size:
5/8

End Condition
Up To Next

Select on the left half of the bottom horizontal line of the part as shown to set the pattern direction. After you select the line, you should see a large grey arrow on the origin pointing to the right, the positive X-direction.

If the arrow is pointing to the left, change the direction of the pattern by clicking the **Reverse Direction** button.

In the **Linear Pattern** PropertyManager, set the **Spacing** to '**3**'.

Set the **Number** to '**7**'.

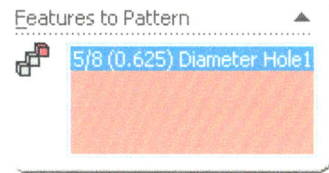

Click in the **Features to Pattern** box. In the flyout FeatureManager design tree, select **5/8 (0.625) Diameter Hole1**.

Click the green check mark button at the top of the **Linear Pattern** PropertyManager.

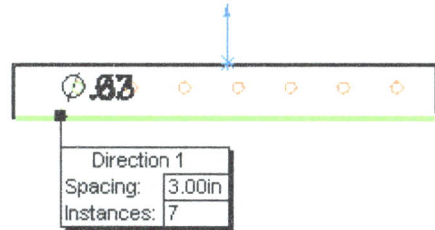

In order to keep this line of holes centered on the end flange, in the FeatureManager design tree, right click on **Annotations** and pick **Show Feature Dimensions**.

Rotate the part by pressing the **Down arrow** key twice so that you can see the **3.00** dimension of the hole spacing. The dimension is between the holes, so you may need to drag it out where it is more visible.

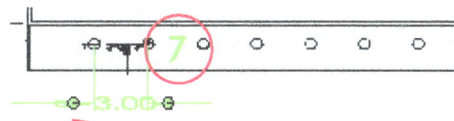

Double click on the **3.00** dimension value to edit it.

In the **Modify** dialog box, pull down the menu and pick **Link Value**.

In the **Shared Values** dialog box, enter the **Name** as '**EndLineSpacing**' and click **OK**.

In the Feature Manager design tree, double click on **LPattern1** to show the **7** dimension. Double click on the **7** dimension. Then, in the **Modify** dialog box, pull down the menu and pick **Link Value**.

In the **Shared Values** dialog box, enter the **Name** as '**EndLineInstances**' and click **OK**.

Since the first hole is placed at the cursor location, you need to edit the sketch in order to locate this hole and the line pattern.

In the FeatureManager design tree, click on the plus sign to the left of **5/8 (0.625) Diameter Hole1** to expand the folder items. There are two sketches listed here. The first sketch contains the hole locations and dimensions. The second sketch is used to create the 3D hole shape.

Right click on the first sketch, **Sketch16**, and pick **Edit Sketch**.

In the bottom left corner of the graphics area, change the View orientation by clicking the pull down arrow and picking **Right**, or press **Ctrl-4**.

Click the **Smart Dimension** icon in the CommandManager, or pull down the "Tools" menu and pick **Dimensions – Smart.**

Create a horizontal dimension from the point to the part origin.

When the **Modify** dialog box appears, pull down the menu and pick **Add Equation**.

In the **Edit Equation** dialog box enter '**EndLineSpacing * (EndLineInstances - 1)/2**'.

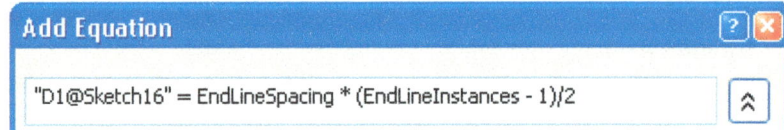

Add Equation

"D1@Sketch16" = EndLineSpacing * (EndLineInstances - 1)/2

Click **OK** twice to close the dialog boxes.

Add a '**1.625**' vertical dimension from the point to the top of the part.

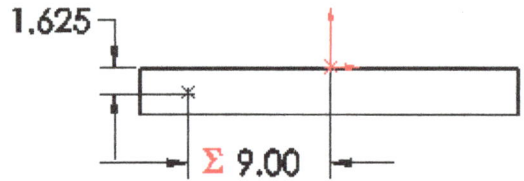

1.625

Σ 9.00

Exit the sketch by clicking the **Exit Sketch** icon in the CommandManager or in the upper right corner of the graphics area.

In the FeatureManager design tree, right click on **Annotations** and pick **Show Feature Dimensions** to hide the dimensions.

Creating the Bolt Hole Circle

Select the right end flange just above the first hole in the line pattern as shown for the plane to sketch on. This is where the new hole will be located.

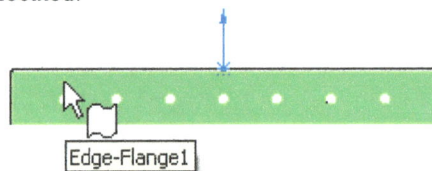

Edge-Flange1

Click the **Hole Wizard** icon in the CommandManager, or pull down the "Insert" menu and pick **Features – Hole – Wizard**.

In the **Hole Specification** PropertyManager, under the **Type** tab, click the **Hole** button.

From the **Size** pull down list select **3/32**.

Under **End Condition**, pick **Up to Next** from the pull down list.

Next, click on the **Positions** tab.

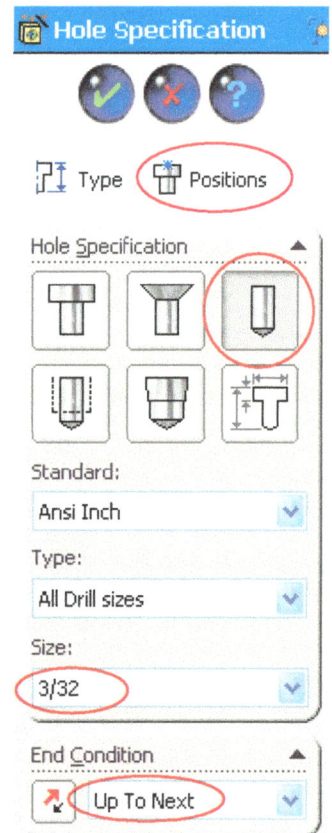

Hole Specification

Type   Positions

Hole Specification

Standard:
Ansi Inch

Type:
All Drill sizes

Size:
3/32

End Condition
Up To Next

Now, zoom in to the left most holes by placing the cursor between the two holes and scrolling the middle mouse button. You may also click the **Zoom To Area** icon from the "View" toolbar. Then, make a box around the area that you want to zoom in to.

Create a vertical centerline line through the center of the 5/8 diameter circle by clicking the **Centerline** icon in the CommandManager, or pull down the "Tools" menu and pick **Sketch Entities – Centerline**.

Press the **Escape** key to deselect the **Centerline** tool.

Select the centerpoint of the 3/32 diameter circle. Then, hold down the **Ctrl** key and select the centerline.

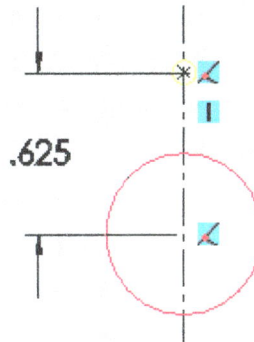

In the **Properties** PropertyManager, under **Add Relations**, click the **Coincident** button.

Create a '**.625**' vertical dimension from the centers of the circles using the **Smart Dimension** icon in the CommandManager, or pull down the "Tools" menu and pick **Dimensions – Smart**.

Click the green check mark button at the top of the **Dimension** PropertyManager and at the top of the **Hole Specification** PropertyManager to accept the settings and create the hole.

Pull down the "View" menu and pick **Temporary Axes** to show any existing axes. A small blue plus sign should now appear in the center of each hole. These are the temporary axes.

Click the **Circular Pattern** icon in the CommandManager, or pull down the "Insert" menu and pick **Pattern/Mirror – Circular Pattern**.

At the top of the **Circular Pattern** PropertyManager, the **Pattern Axis** window is already highlighted. Select the temporary axis through the center of the larger 5/8 diameter circle.

Set the **Number of Instances** to '**8**' and make certain the **Equal spacing** check box is checked.

If **3/32 (0.09375) Diameter Hole1** is not already in the **Features to Pattern** box, click in the box and select **3/32 (0.09375) Diameter Hole1** from the flyout FeatureManager design tree.

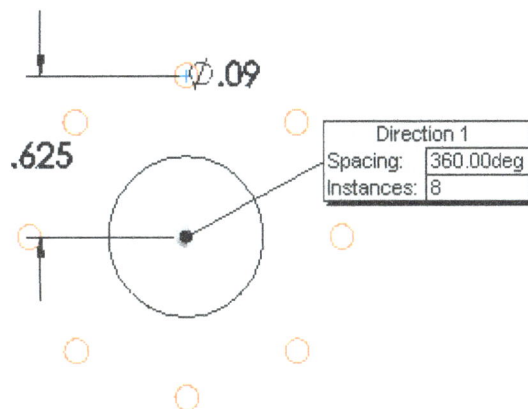

Click the green check mark button to accept the settings and create the circular pattern.

Press the **f** key on the keyboard to **Zoom to Fit**.

Pull down the "View" menu and pick **Temporary Axes** to hide the temporary axes.

Chapter 10: Front Panel   149

Create a Line of Bolt Hole Circles

Click the **Linear Pattern** icon in the CommandManager, or pull down the "Insert" menu and pick **Pattern/Mirror – Linear Pattern**.

Select on the left half of the bottom horizontal line of the part as shown to set the pattern direction.

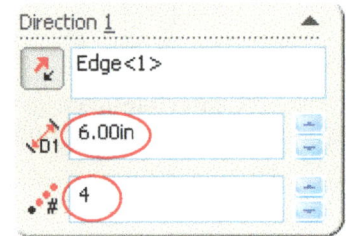

In the **Linear Pattern** PropertyManager, set the **Spacing** to '6'.

Set the **Number** to '4'.

If **CirPattern1** is not already in the **Features to Pattern** box, click in the box and select **CirPattern1** from the flyout FeatureManager design tree.

Click the green check mark button at the top of the **Linear Pattern** PropertyManager.

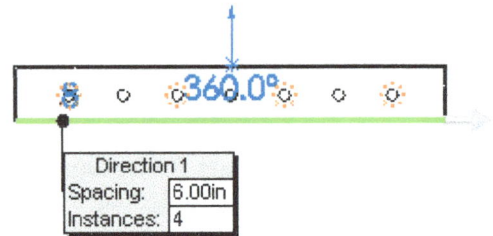

To keep the line of bolt hole circles aligned with the other holes, in the FeatureManager design tree, right click on **Annotations** and pick **Show Feature Dimensions**.

Rotate the part by pressing the **Down arrow** key twice so that you can see the **6.00** dimension of the hole spacing. The dimension is between the holes, so you may need to drag it down.

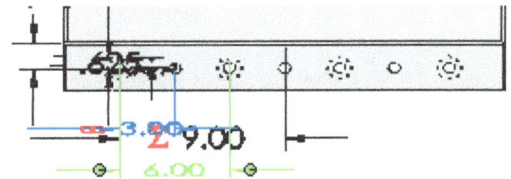

Double click on the **6.00** dimension to edit it.

In the **Modify** dialog box, pull down the menu and pick **Add Equation**.

In the **Edit Equation** dialog box, type '**2*EndLineSpacing**'.

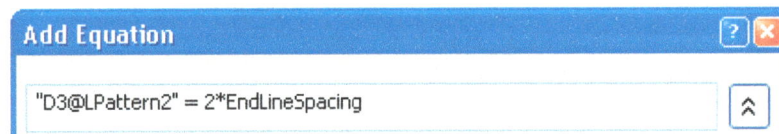

"D3@LPattern2" = 2*EndLineSpacing

Click **OK** twice to close the dialog boxes

In the FeatureManager design tree, right click on **Annotations** and pick **Show Feature Dimensions** to hide the dimensions.

Mirror the Holes to the Other End Flange

Click the **Mirror** icon in the CommandManager, or pull down the "Insert" menu and pick **Pattern/Mirror – Mirror**.

In the flyout FeatureManager design tree, select **Right Plane** as the **Mirror Face/Plane**.

Under **Features to Mirror**, select **LPattern1** and **LPattern2** from the flyout FeatureManager design tree.

Click the green check mark button at the top of the **Mirror** PropertyManager.

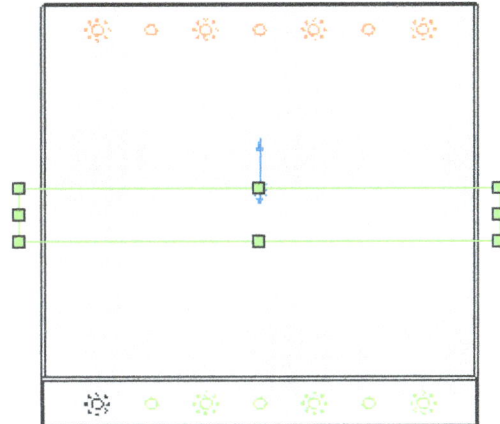

Holes on the Front Flange

In the bottom left corner of the graphics area, change the View orientation by clicking the pull down arrow and picking **Front**, or press **Ctrl-1**.

Select the front edge flange toward the left end for the plane to sketch on as shown.

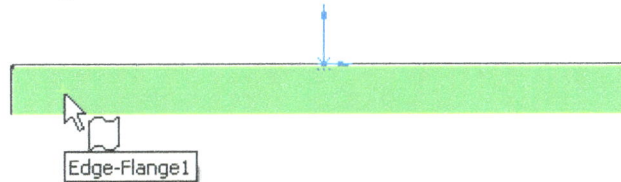

Edge-Flange1

Click the **Hole Wizard** icon in the CommandManager, or pull down the "Insert" menu and pick **Features – Hole – Wizard**.

In the **Hole Specification** PropertyManager, under the **Type** tab, click the **Hole** button.

From the **Size** pull down list, select **3/4**.

Under **End Condition**, pick **Up to Next**.

Next, click on the **Positions** tab.

Click along the front plane to place four more holes as shown.

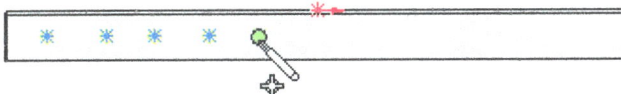

Press the **Escape** key to deselect the **Point** tool.

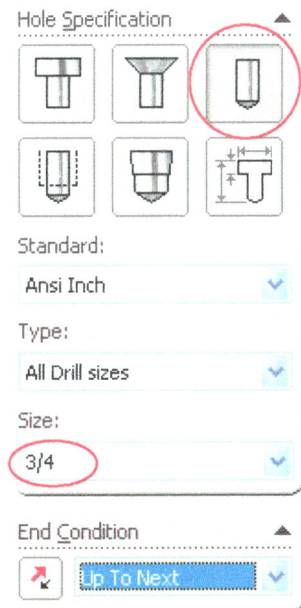

Hole Specification

Standard:
Ansi Inch

Type:
All Drill sizes

Size:
3/4

End Condition
Up To Next

With the centerpoint of the rightmost circle selected, hold down the **Ctrl** key and select the origin.

In the **Properties** PropertyManager, click the **Vertical** button.

Select the centerpoint of the leftmost circle. **Ctrl** select the other four centerpoints. Or you can create a window around them to select them.

In the **Properties** PropertyManager, click the **Horizontal** button.

Click the **Smart Dimension** icon in the CommandManager, or pull down the "Tools" menu and pick **Dimensions – Smart.**

Create a '**3**' horizontal dimension from the leftmost centerpoint to the outside left edge of the part. Then, place a '**2**' horizontal dimension between the leftmost centerpoint and the next centerpoint. You may need to zoom in to dimension the correct points.

Create a '**1.75**' vertical dimension from the leftmost centerpoint to the top of the part as shown.

Create a '**2.25**' horizontal dimension between the rightmost centerpoint and the next centerpoint to the left as shown below.

Double click on the **2.25** dimension. Then, in the **Modify** dialog box, pull down the menu and pick **Link Value.**

In the **Shared Values** dialog box, enter the **Name** as '**FrontLineSpacing**' and click **OK**.

Place a horizontal dimension between the last centerpoint used and the next centerpoint to the left as shown. Then, in the **Modify** dialog box, pull down the menu and pick **Link Value.**

In the **Shared Values** dialog box, pull down the menu and pick '**FrontLineSpacing**'. This ensures that the two dimensions will always be the same value.

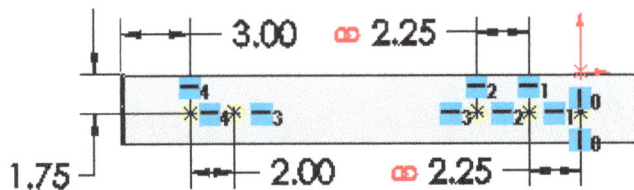

Create a vertical centerline line from the part origin to the bottom of the flange by clicking the **Centerline** icon in the CommandManager, or pull down the "Tools" menu and pick **Sketch Entities – Centerline**.

Press the **Escape** key to deselect the centerline.

Click the **Mirror Entities** icon from the CommandManager, or pull down the "Tools" menu and pick **Sketch Tools – Mirror**.

In the **Mirror** PropertyManager, under **Entities to mirror**, select the four centerpoints to the left of the centerline.

Under **Mirror about**, select the vertical centerline.

Click the green check mark button at the top of the **Mirror** PropertyManager, and then at the top of the **Hole Position** PropertyManager to create the holes.

The Back Flange

In the bottom left corner of the graphics area, change the View orientation by clicking the pull down arrow and picking **Back**, or press **Ctrl-2**.

Select the back edge flange under the origin marker as the plane to sketch on.

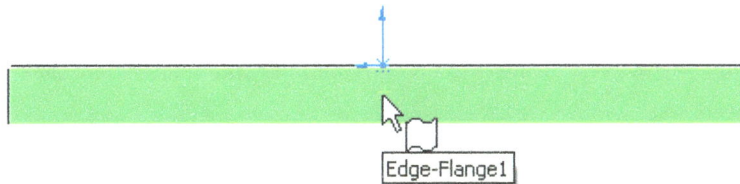

Edge-Flange1

Click the **Hole Wizard** icon in the CommandManager, or pull down the "Insert" menu and pick **Features – Hole – Wizard**.

Set the **End Condition** to **Up to Next**.

Then, click on the **Positions** tab.

Place six more holes by picking along the flange as shown.

Press the **Escape** key to deselect the **Point** tool.

Ctrl select the centerpoints of the middle hole and the two outer holes. Then, in the **Properties** PropertyManager, click the **Horizontal** button.

Ctrl select the centerpoints of the other four holes. Then, in the **Properties** PropertyManager, click the **Horizontal** button. All seven holes will not be at the same height.

Ctrl select the centerpoint of the center hole and the origin. Then, in the **Properties** PropertyManager, click the **Vertical** button.

Create two vertical centerlines between the second and third hole from each end as shown by clicking the **Centerline** icon in the CommandManager, or pull down the "Tools" menu and pick **Sketch Entities – Centerline**.

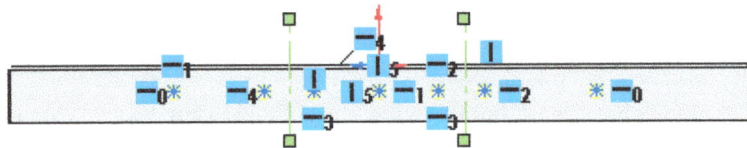

Press the **Escape** key to deselect the **Centerline** tool.

Click the **Smart Dimension** icon in the CommandManager, or pull down the "Tools" menu and pick **Dimensions – Smart.**

Add a '**1**' horizontal dimension from the left centerline to the centerpoint of the hole on the left of the centerline.

Double click the 1.00 dimension. Then, in the **Modify** dialog box, pull down the menu and pick **Link Value**.

In the **Shared Values** dialog box, enter the **Name** as '**BackSpacing1**' and click **OK**.

Place a dimension from the left centerline to the centerpoint of the hole on the right of the centerline. Then, in the **Modify** dialog box, pull down the menu and pick **Link Value**.

In the **Shared Values** dialog box, pull down the menu and pick **BackSpacing1** and click **OK**.

Repeat this step on the two holes around the right centerline.

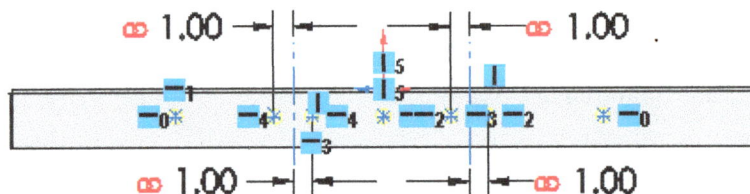

Press **Delete** and then click on the green check mark in the upper right corner of the graphics area. Next, in the FeatureManager design tree, right click on the **Equations** folder and pick **Add Equation**.

In the **Add Equation** dialog box, enter '**BackSpacing2 = BackSpacing1 * 5**' and click **OK** twice to exit the dialog boxes.

In the FeatureManager design tree, click the small plus sign next to **3/4 (0.75) Diameter Hole2** to expand the feature. Then, right click on **Sketch22** and pick **Edit Sketch**.

Place a horizontal dimension from the left centerline to the centerpoint of the center hole. Then, in the **Modify** dialog box, pull down the menu and pick **Link Value**.

In the **Shared Values** dialog box, pull down the menu and pick '**$VAR: BackSpacing2** and click **OK**.

Repeat the above steps to add a dimension between the right centerline and the centerpoint of the center hole, and between the centerpoint of the two outside holes and the appropriate centerlines.

Finally, add a '**.5**' vertical dimension from the bottom of the part to the leftmost centerpoint and a '**1.25**' vertical dimension from the bottom of the part to the centerpoint second from the left.

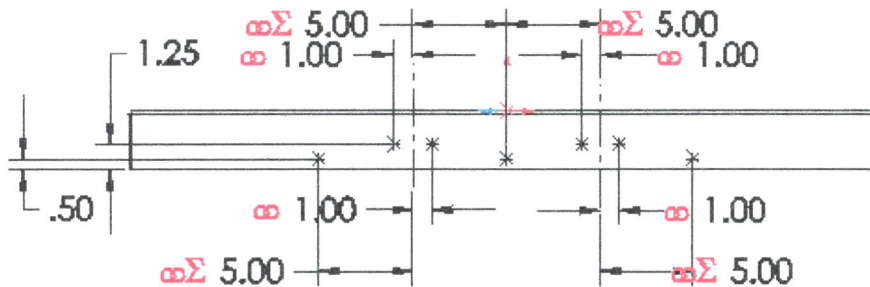

Exit the sketch by clicking the **Exit Sketch** icon in the CommandManager or in the upper right corner of the graphics area.

Select the back edge flange near the lower left corner.

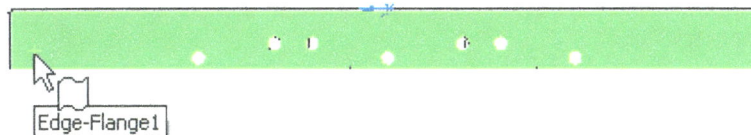

Click the **Hole Wizard** icon in the CommandManager, or pull down the "Insert" menu and pick **Features – Hole – Wizard**.

Set the **Size** to **1/4** and the **End Condition** to **Up to Next**.

Then, click on the **Positions** tab.

Place six more holes (two above each of the three lower 3/4" diameter holes) by picking along the flange as shown.

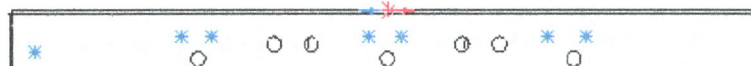

Press the **Escape** key to deselect the **Point** tool.

Pull down the "View" menu and pick **Temporary Axes** to show existing axes.

Ctrl select the centerpoint of the leftmost 1/4" diameter hole and the center (temporary axis) of one of the lower 3/4" diameter holes. Then, in the **Properties** PropertyManager, click the **Horizontal** button.

Ctrl select the remaining six centerpoints of the 1/4" diameter holes and the temporary axis of one of the upper 3/4" diameter holes. Then, in the **Properties** PropertyManager, click the **Horizontal** button.

Click the **Smart Dimension** icon in the CommandManager, or pull down the "Tools" menu and pick **Dimensions – Smart.**

Create a '**4**' horizontal dimension from the leftmost centerpoint of the 1/4" diameter hole to the left most 3/4" diameter hole.

Place a horizontal dimension for the upper leftmost centerpoint to the 3/4" hole.

In the **Modify** dialog box, pull down the menu and pick **Link Value**. Then, in the **Shared Values** dialog box, pull down the menu and select **BackSpacing1**.

Do this for all six upper centerpoints to the temporary axis of the appropriate 3/4" hole.

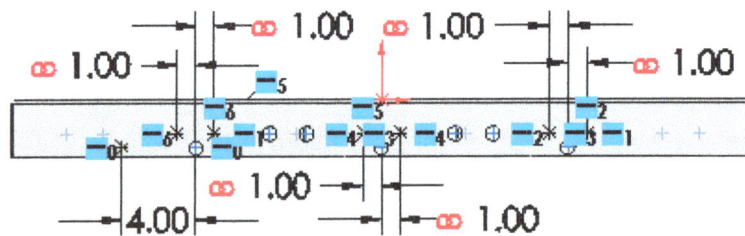

Pull down the "View" menu and pick **Temporary Axes** to hide the temporary axes.

Now, create a linear sketch pattern by pulling down the "Tools" menu and picking **Sketch Tools – Linear Pattern**.

In the graphics area, select the leftmost centerpoint for the **Entities to Pattern**.

Then, under **Direction 1**, set the **Spacing** to '**.99**'.

Check the **Add dimension** check box and set the **Number of Instances** to '**29**'.

At the bottom of the **Linear Pattern** PropertyManager, click on the arrow to expand **Instances to Skip**.

SolidWorks for the Sheet Metal Guy

Click in the **Instances to Skip** box and pick the fourth (4,1), fifth (5,1), and sixth (6,1) instances of the line pattern counting from the left.

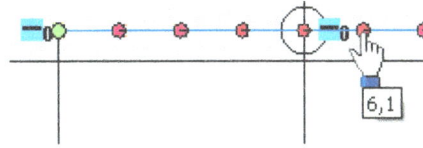

Continue picking the following instances to skip: 9, 10, 11, 14, 15, 16, 19, 20, 21, 24, 25, and 26.

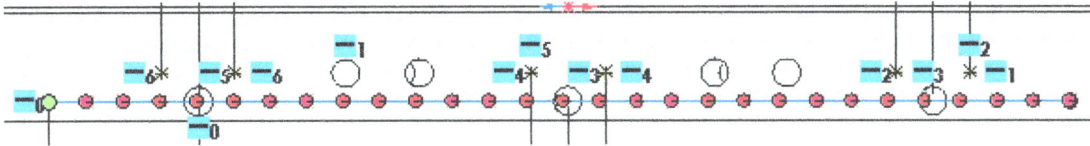

Click the green check mark button at the top of the **Linear Pattern** PropertyManager to accept the settings and create the line of holes.

Double click on the **.99** dimension of the line spacing.

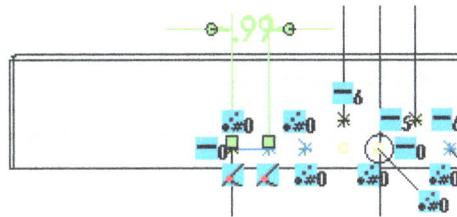

In the **Modify** dialog box, pull down the menu and pick **Link Value**. Then, in the **Shared Values** dialog box, pull down the menu and select **BackSpacing1**.

In the upper right corner of the graphics area, click the green check mark to close the Hole Wizard.

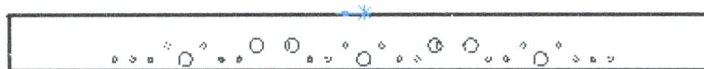

Create a Cutout

In the bottom left corner of the graphics area, change the View orientation by clicking the pull down arrow and picking **Top**, or press **Ctrl-5**.

Click the **Extruded Cut** icon in the CommandManager, or pull down the "Insert" menu and pick **Cut – Extrude**.

Pick anywhere on the top of the part when prompted to select a plane on which to sketch the feature cross-section.

□ Create a rectangle below the origin using the **Rectangle** icon in the CommandManager, or pull down the "Tools" menu and pick **Sketch Entities – Rectangle**.

⊕ Click the **Circle** icon in the CommandManager, or pull down the "Tools" menu and pick **Sketch Entities – Circle**.

Place a circle on all four corners of the rectangle.

Press the **Escape** key to cancel the active command.

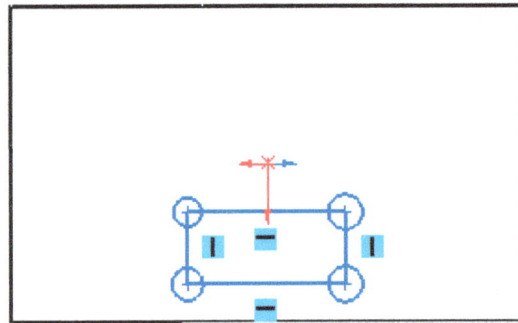

= Ctrl select the four circles and in the **Properties** PropertyManager, click the **Equal** button.

✎ Add a '**.25**' diameter dimension to one of the circles using the **Smart Dimension** icon in the CommandManager, or pull down the "Tools" menu and pick **Dimensions – Smart**

Pull down the "Tools" menu and pick **Sketch Tools – Chamfer**.

In the **Sketch Chamfer** dialog box, click the **Angle-distance** radio button.

Set the **Distance 1** to '**1**' and **Distance 1 Angle** to '**45**'.

Pick each of the four corners of the rectangle to chamfer the corners.

✔ Click the green check mark button to close the Sketch Chamfer PropertyManager.

Select the four **1.00** dimensions created by the **Chamfer** command and delete them.

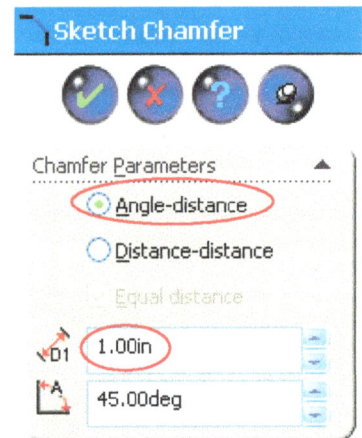

= Ctrl select the four chamfer lines and in the **Properties** PropertyManager, click the **Equal** button.

✎ Click the **Smart Dimension** icon in the CommandManager, or pull down the "Tools" menu and pick **Dimensions – Smart.**

Dimension one of the chamfer lines to be '**1**' in length.

Add a '**10**' horizontal dimension to the rectangle.

Add a '**2**' vertical dimension to one of the vertical lines of the rectangle.

Right click on one of the vertical lines and pick **Select Midpoint**. Move the cursor to the bottom edge of the part and pick the horizontal edge line to create a '**4.75**' vertical dimension from the edge of the part to the center of the cutout.

Press the **Escape** key to deselect the active command.

Right click on the top horizontal line of the rectangle and pick **Select Midpoint**. Hold down the **Ctrl** key and select the origin.

In the **Properties** PropertyManager, click the **Vertical** button.

Exit the sketch by clicking the **Exit Sketch** icon in the CommandManager or in the upper right corner of the graphics area.

In the **Cut-Extrude** PropertyManager, check the **Link to thickness** check box and then click the green check mark button at the top of the **Cut-Extrude** PropertyManager.

Create a Herring Bone Pattern

Click the **Sheet Metal** icon in the control area of the CommandManager. Then, click the **Extruded Cut** icon from the toolbar, or pull down the "Insert" menu and pick **Cut – Extrude**.

Pick anywhere on the top of the part when prompted to select a plane on which to sketch the feature cross-section.

Create a horizontal centerline from the centerpoint of the left vertical line of the cutout by clicking the **Centerline** icon in the CommandManager, or pull down the "Tools" menu and pick **Sketch Entities – Centerline**.

Click the **Circle** icon in the CommandManager, or pull down the "Tools" menu and pick **Sketch Entities – Circle**.

Place a circle on the left end of the centerline.

Create two more circles, one above the first and the other to the left of the first. Use the dotted lines to help get the circles horizontal and vertical to each other.

Press the **Escape** key to deselect the **Circle** tool.

Ctrl select the three circles. Then, in the **Properties** PropertyManager, click the **Equal** button.

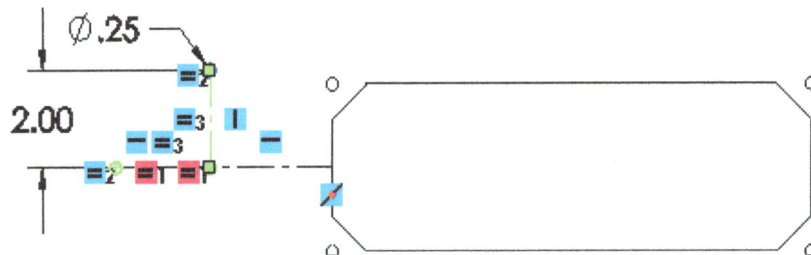

Click the **Centerline** icon in the CommandManager, or pull down the "Tools" menu and pick **Sketch Entities – Centerline**.

Create a horizontal centerline from the center of the leftmost circle to the endpoint of the first centerline. Then, create a vertical centerline to the center of the top circle.

Add a '**0.25**' diameter dimension to the top circle using the **Smart Dimension** icon in the CommandManager, or pull down the "Tools" menu and pick **Dimensions – Smart.**

Then, add a '**2**'dimension between the two circles as shown.

Press the **Escape** key and **Ctrl** select the two centerlines created in the previous step.

In the **Properties** PropertyManager, click the **Equal** button.

Create two rectangles on top of the centerlines using the **Rectangle** icon in the CommandManager, or pull down the "Tools" menu and pick **Sketch Entities – Rectangle**.

Create a construction line diagonally across each of the rectangles using the **Centerline** icon in the CommandManager, or pull down the "Tools" menu and pick **Sketch Entities – Centerline**.

Press the **Escape** key to deselect the **Centerline** tool.

Right click on one of the diagonal centerlines and pick **Select Midpoint**. Then, **Ctrl** select the centerline which passes through the rectangle.

In the **Properties** PropertyManager, under **Add Relations**, click the **Midpoint** button.

Repeat the above step for the other rectangle.

Select the two long sides of both rectangles. Then, in the **Properties** PropertyManager, click the **Equal** button.

Next, select the two short sides of both rectangles. Then, in the **Properties** PropertyManager, click the **Equal** button.

Dimension one of the rectangles to be '**1**' by '**.25**' using the **Smart Dimension** icon in the CommandManager, or pull down the "Tools" menu and pick **Dimensions – Smart.**

Next, add a '**1.5**' dimension from the center of the lower right circle to the large rectangular cutout.

Press the **Escape** key to deselect the active command.

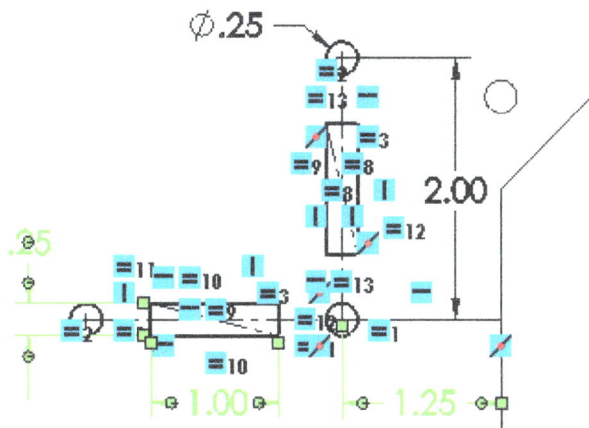

Window select all of the sketch entities. Then, hold down the **Ctrl** key and select the horizontal centerline which ties the pattern to the cutout. This should remove (un-highlight) this line from the selection list.

Pull down the "Tools" menu and pick **Sketch Tools – Rotate**.

In the graphics area, select the center of the lower right circle as the **Base point**.

In the **Rotate** PropertyManager, uncheck the **Keep relations** check box.

Under **Center of rotation**, set the **Angle** to '**45**'.

Click the green check mark button at the top of the **Rotate** PropertyManager.

Exit the sketch by clicking the **Exit Sketch** icon in the CommandManager or in the upper right corner of the graphics area.

In the **Cut-Extrude** PropertyManager, check the **Link to thickness** check box, and then, click the green check mark button at the top of the **Cut-Extrude** PropertyManager.

In the FeatureManager design tree, slowly click two times (left click once, pause, left click again) on **Cut-Extrude2** to select it. Then, enter '**Herring Bone**' for the new name.

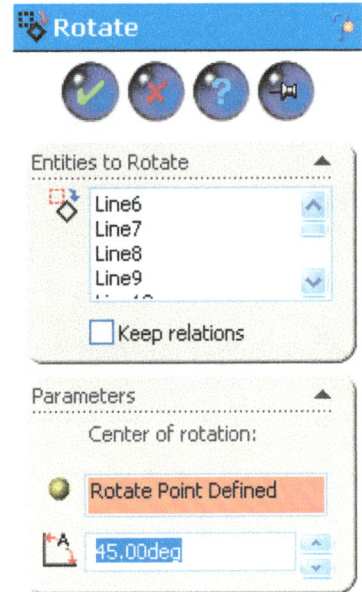

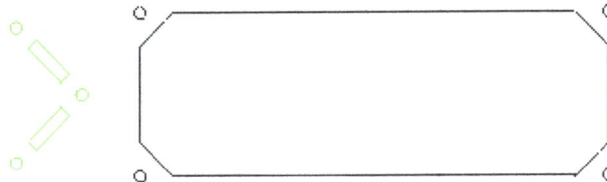

Create a Line Pattern

Click the **Features** icon in the control area of the CommandManager. Then, click the **Linear Pattern** icon in the CommandManager, or pull down the "Insert" menu and pick **Pattern/Mirror – Linear Pattern**.

Select the bottom horizontal line of the part toward the right corner to set the pattern direction.

In the **Linear Pattern** PropertyManager, set the **Spacing** to '**.75**' and the **Number of Instances** to '**8**'.

If **Herring Bone** is not in the **Features to Pattern** box, click in the **Features to Pattern** box, and select **Herring Bone** from the flyout FeatureManager design tree.

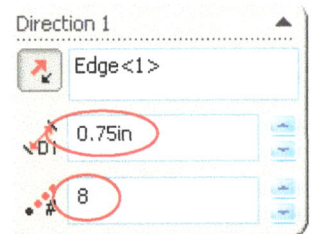

Click the green check mark button at the top of the **Linear Pattern** PropertyManager.

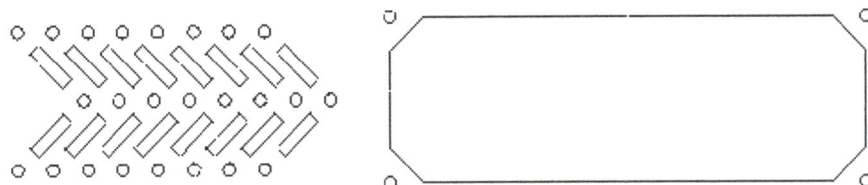

Create a Double D

Click the **Sheet Metal** icon in the control area of the CommandManager. Then, click the **Extruded Cut** icon from the toolbar or pull down the "Insert" menu and pick **Cut – Extrude**.

Pick anywhere on the top of the part when prompted to select a plane on which to sketch the feature cross-section.

Place a circle above the herring bone pattern using the **Circle** icon in the CommandManager, or pull down the "Tools" menu and pick **Sketch Entities – Circle**.

Create two parallel vertical lines inside the circle using the **Line** icon, or pull down the "Tools" menu and pick **Sketch Entities – Line**.

Next, click the **Trim Entities** icon, or pull down the "Tools" menu and pick **Sketch Tools - Trim Entities**.

In the **Trim** PropertyManager, make sure that the **Power Trim** button is selected.

Place the cursor outside the left side of the circle. Press and hold the left mouse button and drag the cursor across the edge of the circle. Do the same on the other side of the circle.

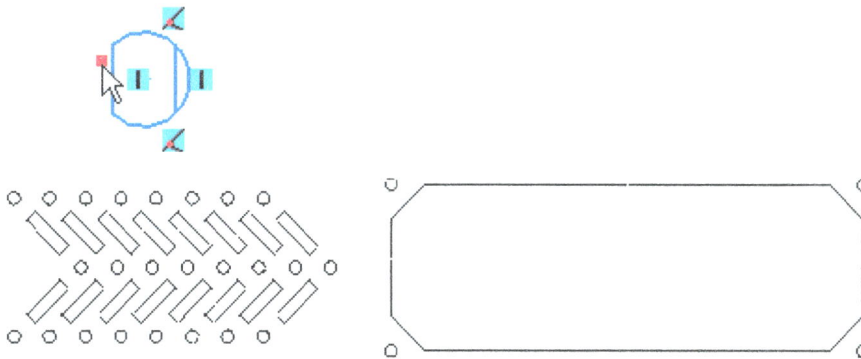

Press the **Escape** key to deselect the active command.

Ctrl select the two vertical lines. Then, in the **Properties** PropertyManager, click the **Equal** button.

Click the **Smart Dimension** icon in the CommandManager, or pull down the "Tools" menu and pick **Dimensions – Smart**.

Place a '.75' horizontal dimension between the two vertical lines. Then, place a '.5' radius dimension to one of the arcs. Next, right click on the 0.5 dimension and pick **Properties** from the menu.

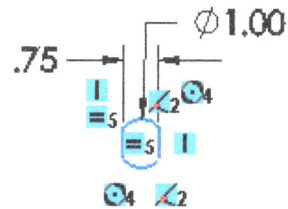

In the **Properties** dialog box, check the **Diameter dimension** check box and click the **OK** button.

Press the **Escape** key to deselect the active command.

Click the **Smart Dimension** icon in the CommandManager, or pull down the "Tools" menu and pick **Dimensions – Smart.**

Locate the Double D shape by adding a '7' horizontal dimension from the origin to the centerpoint of the Double D. Then, add a '9' vertical dimension from the bottom of the part to the centerpoint of the Double D as shown.

Exit the sketch by clicking the **Exit Sketch** icon in the CommandManager or in the upper right corner of the graphics area.

In the **Cut-Extrude** PropertyManager, check the **Link to thickness** check box and then click the green check mark button at the top of the **Cut-Extrude** PropertyManager.

In the FeatureManager design tree, slowly click two times (left click once, pause, left click again) on **Cut-Extrude2** to select it. Then, enter '**Double D**' for the new name.

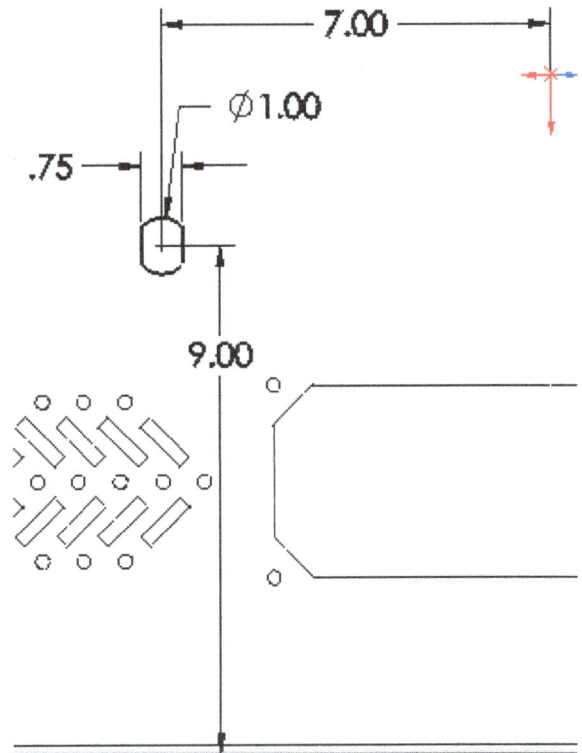

Create a Line Pattern

Click the **Features** icon in the control area of the CommandManager. Then, click the **Linear Pattern** icon in the CommandManager, or pull down the "Insert" menu and pick **Pattern/Mirror – Linear Pattern.**

Select the bottom horizontal line of the part toward the right corner to set the pattern direction as shown.

In the **Linear Pattern** PropertyManager, set the **Spacing** to '5' and the **Number of Instances** to '2'.

If **Double D** is not in the **Features to Pattern** box, click in the **Features to Pattern** box, and select **Double D** from the flyout FeatureManager design tree.

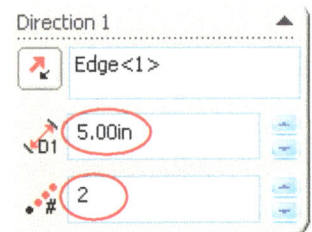

Click the green check mark button at the top of the **Linear Pattern** PropertyManager.

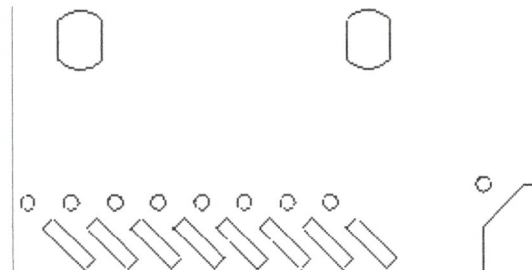

Create a Diamond Shape

Click the **Extruded Cut** in the CommandManager, or pull down the "Insert" menu and pick **Cut – Extrude**.

Pick anywhere on the top of the part when prompted to select a plane on which to sketch the feature cross-section.

Create a rectangle near the upper left corner of the part using the **Rectangle** icon in the CommandManager, or pull down the "Tools" menu and pick **Sketch Entities – Rectangle**.

Click the **Add Relation** icon in the CommandManager, or pull down the "Tools" menu and pick **Relations – Add**.

In the **Properties** PropertyManager, click the **Equal** button. Now it is a square.

Add a '**.75**' dimension to the left side of the rectangle using the **Smart Dimension** icon in the CommandManager, or pull down the "Tools" menu and pick **Dimensions – Smart**.

Pull down the "Tools" menu and pick **Sketch Tools – Rotate**.

In the graphics area, select the four lines of the rectangle.

In the **Rotate** PropertyManager, under **Parameters**, click in the **Base point** box.

Then, in the graphics area, pick inside of the square to indicate the **Center of Rotation**.

In the **Rotate** PropertyManager, set the **Angle** to '**45**'.

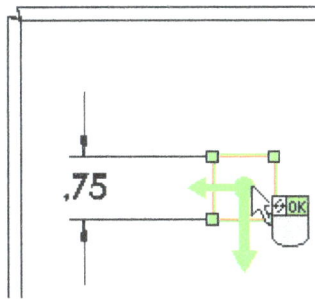

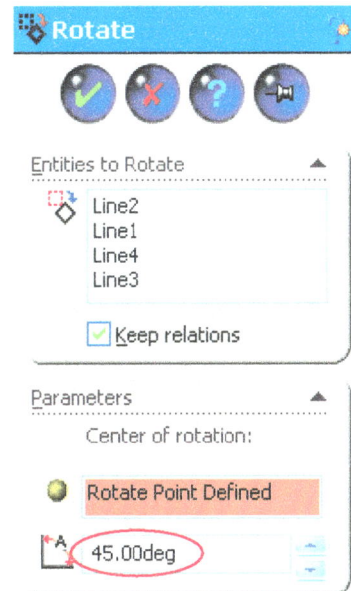

Click the green check mark button at the top of the **Rotate** PropertyManager.

Ctrl select the top two lines of the diamond. Then, in the **Properties** PropertyManager, click the **Perpendicular** button.

Create a horizontal centerline below the diamond by clicking the **Centerline** icon in the CommandManager, or pull down the "Tools" menu and pick **Sketch Entities – Centerline**.

Click the **Smart Dimension** icon in the CommandManager, or pull down the "Tools" menu and pick **Dimensions – Smart**.

Add a '45' angular dimension from the lower right angular line of the diamond and the centerline. Add a '2.5' vertical dimension from the left side point of the diamond to the centerline. Then, add a '5' vertical dimension from the centerline to the origin. Next, add a '15' horizontal dimension from the top of the diamond to the origin.

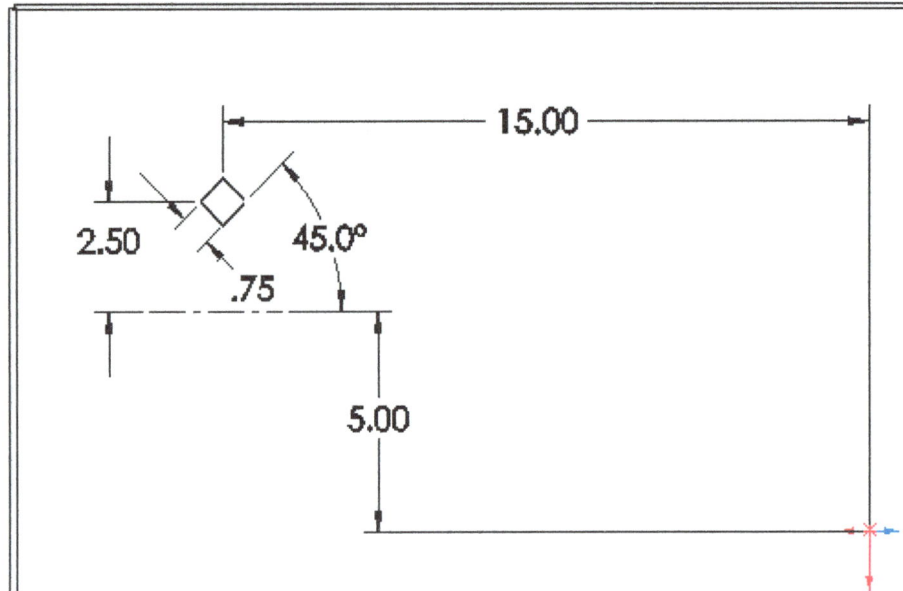

Exit the sketch by clicking the **Exit Sketch** icon in the CommandManager or in the upper right corner of the graphics area.

In the **Cut-Extrude** PropertyManager, check the **Link to thickness** check box and then click the green check mark button at the top of the **Cut-Extrude** PropertyManager.

In the FeatureManager design tree, slowly click two times (left click once, pause, left click again) on **Cut-Extrude2** to select it. Then, enter '.75 Square at 45' for the new name.

Create a Keyhole

Click the **Extruded Cut** in the CommandManager, or pull down the "Insert" menu and pick **Cut – Extrude**.

Pick anywhere on the top of the part when prompted to select a plane on which to sketch the feature cross-section.

Click the **Circle** icon in the CommandManager, or pull down the "Tools" menu and pick **Sketch Entities – Circle**.

Place a circle to the right of the diamond. Place a smaller circle at the bottom quadrant point of the first circle as shown.

Add a '.25' diameter dimension the smaller circle to be and a '.625' diameter dimension to the larger circle using the **Smart Dimension** icon in the CommandManager, or pull down the "Tools" menu and pick **Dimensions – Smart.**

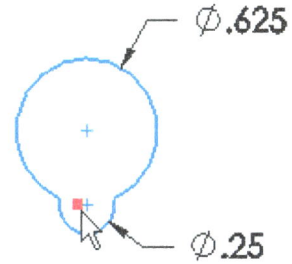

Next, click the **Trim Entities** icon, or pull down the "Tools" menu and pick **Sketch Tools - Trim Entities**.

In the **Trim** PropertyManager, make sure that the **Power Trim** button is selected. Place the cursor inside the larger circle. Press and hold the left mouse button and drag the cursor across the edge of the smaller circle and the larger circle creating a single keyhole shape.

Click on the **3 Point Arc** icon in the CommandManager, or pull down the "Tools" menu and pick **Sketch Entities – 3 Point Arc**.

3 Point Arc
Sketches a 3 point arc. Select start and end points, then drag the arc to set the radius or to reverse the arc.

Click on the centerpoint of the large circle. Next, move the cursor down below the keyhole and click. Then, move the cursor to the left and click to create an arc.

Press the **Escape** key.

Ctrl select the three points that you used to create the arc. Then, in the **Properties** PropertyManager, click the **Vertical** button.

In the graphics area, select the arc.

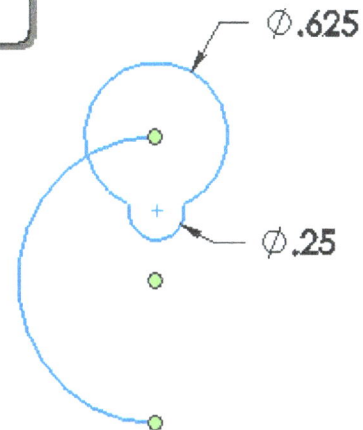

In the **Arc** PropertyManager, under **Options**, check the **For construction** check box. This changes the arc into a centerline.

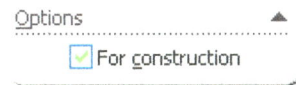

Click the **Smart Dimension** icon in the CommandManager, or pull down the "Tools" menu and pick **Dimensions – Smart**.

Add a '5' horizontal dimension from the centerpoint of the keyhole to the origin. Then, add a '2.75' vertical dimension from the centerpoint of the large circle to the centerpoint of the arc. Also, add a '5' vertical dimension from the origin to the centerpoint of the arc as shown.

Exit the sketch by clicking the **Exit Sketch** icon in the CommandManager or in the upper right corner of the graphics area.

In the **Cut-Extrude** PropertyManager, check the **Link to thickness** check box and then click the green check mark button at the top of the Cut-Extrude PropertyManager.

In the FeatureManager design tree, slowly click two times (left click once, pause, left click again) on **Cut-Extrude3** to select it. Then, enter '0.625 Keyhole' for the new name.

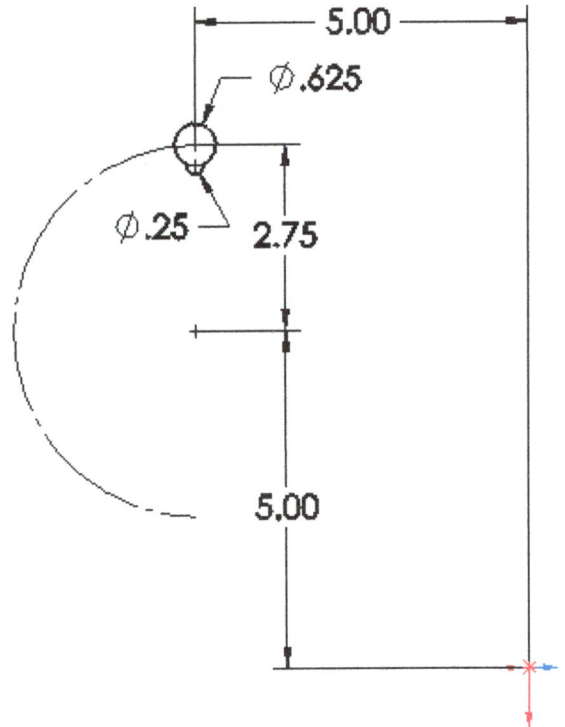

Create a Curve Driven Pattern

In the FeatureManager design tree, click on the plus to the left of **0.625 Keyhole**. Right click on the sketch and pick **Show** from the menu. This displays the sketch entities even though you are not editing the sketch.

Pull down the "Insert" menu and pick **Pattern/Mirror – Curve Driven Pattern**.

In the graphics area, select the arc centerline as the curve to pattern along.

In the **Curve Driven Pattern** PropertyManager, under **Direction 1**, set the **Number of Instances** to '4'.

Check the **Equal spacing** check box.

Make sure that the **Transform curve** and **Tangent to curve** radio buttons are selected.

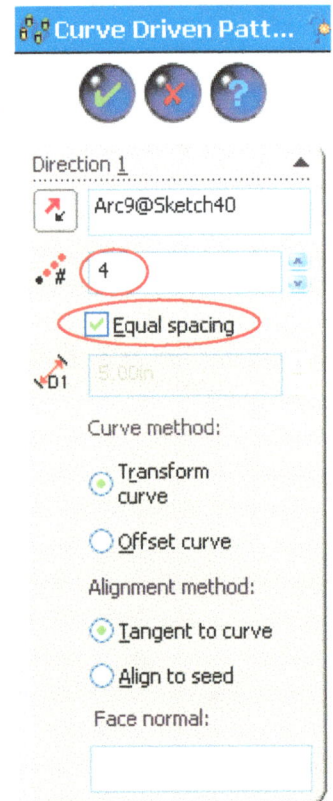

Click in the **Features to Pattern** box. In the flyout FeatureManager design tree, select **0.625 Keyhole**.

Note that the keyhole shapes rotate as they are copied around the arc. Should you want the shapes to maintain their original orientation, under **Direction 1**, pick the **Align to seed** radio button.

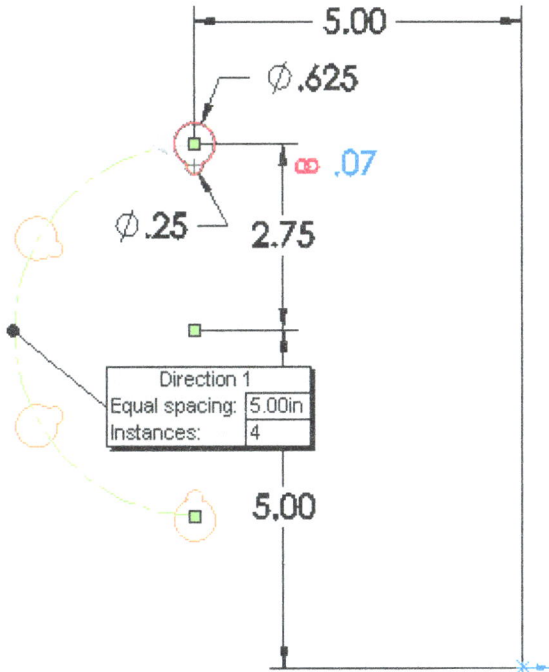

Click the green check mark button at the top of the **Curve Driven Pattern** PropertyManager.

In the FeatureManager design tree, right click on the sketch under **0.625 Keyhole** and pick **Hide**.

Mirror the Holes

Click the **Mirror** icon in the CommandManager, or pull down the "Insert" menu and pick **Pattern/Mirror – Mirror**.

In the flyout FeatureManager design tree, select **Right Plane** as the **Mirror Face/Plane**.

Under **Features to Mirror**, select **LPattern3, LPattern4, .75 Square at 45**, and **CrvPattern1** from the flyout FeatureManager design tree.

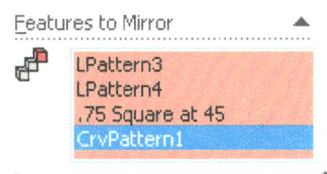

Click the green check mark button at the top of the **Mirror** PropertyManager.

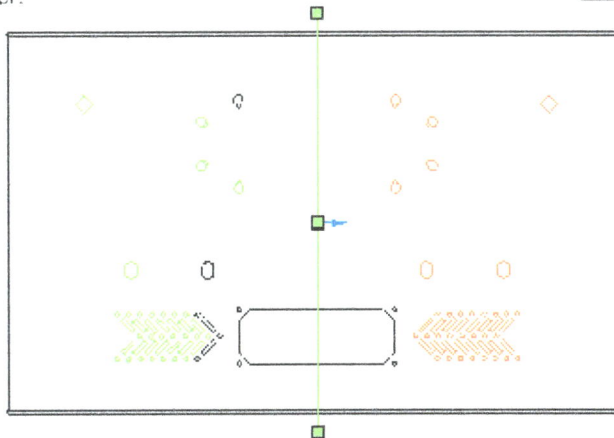

Create a Hexagon

Click the **Extruded Cut** icon in the CommandManager, or pull down the "Insert" menu and pick **Cut – Extrude**.

Pick anywhere on the top surface of the part when prompted to select a plane on which to sketch the feature cross-section.

Pull down the "Tools" menu and pick **Sketch Entities – Polygon**.

In the **Polygon** PropertyManager, set the **Number of Sides** to '**6**'.

Pick on the left side of the part to indicate the location of the hexagon. As you move to the left, notice that the shape is also rotating with the cursor movement.

Before picking a second time to set the size, make certain that the orientation is at **0** degrees. This places a flat side horizontally at the bottom of the hexagon.

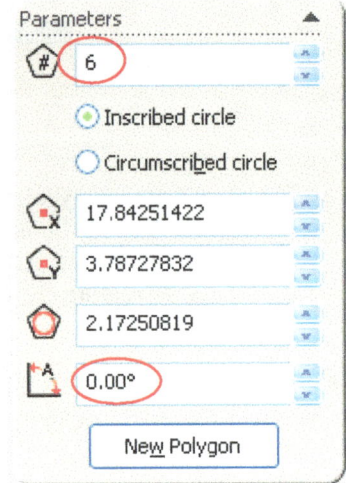

Click the **Centerline** icon in the CommandManager, or pull down the "Tools" menu and pick **Sketch Entities – Centerline**.

Create a vertical centerline starting at the centerpoint of the hexagon and going up. Then, create an angled centerline to the right and the left starting at the centerpoint of the hexagon.

Click the **Smart Dimension** icon in the CommandManager, or pull down the "Tools" menu and pick **Dimensions – Smart**.

Create a '**.25**' horizontal dimension across the width of the hexagon from the left corner point to the right corner point. Next, add a '**3.5**' horizontal dimension from the center of the hexagon to the left edge of the part. Then, add '**4.75**' vertical dimension from the center of the hexagon to the bottom of the part. Finally, add two '**27.5**' angular dimensions between the vertical centerline and the angled centerlines as shown.

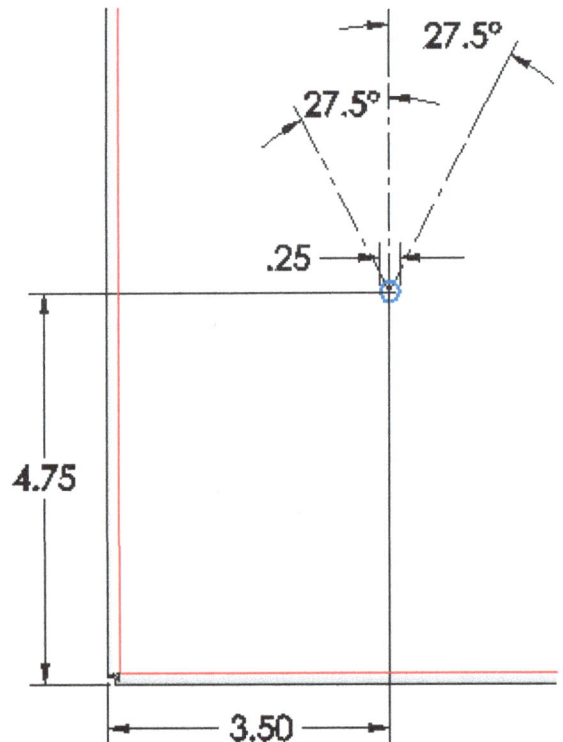

Exit the sketch by clicking the **Exit Sketch** icon in the CommandManager or in the upper right corner of the graphics area.

In the **Cut-Extrude** PropertyManager, check the **Link to thickness** check box and then click the green check mark button at the top of the **Cut-Extrude** PropertyManager.

In the FeatureManager design tree, slowly click two times (left click once, pause, left click again) on **Cut-Extrude4** to select it. Then, enter '**0.25 Hexagon**' for the new name.

In the FeatureManager design tree, click on the plus to the left of **0.25 Hexagon**. Right click on the sketch and pick **Show**.

A Grid of Hexagons

Click the **Linear Pattern** icon in the CommandManager, or pull down the "Insert" menu and pick **Pattern/Mirror – Linear Pattern**.

In the graphics area, select the bottom half of the angled centerline on the right to set the pattern direction for **Direction 1**.

Then, select the bottom half of the angled centerline on the left to set the pattern direction for **Direction 2**.

In the **Linear Pattern** PropertyManager, under **Direction 1**, set the **Spacing** to '**.5625**' and the **Number of Instances** to '**7**'.

In the **Linear Pattern** PropertyManager, under **Direction 2**, set the **Spacing** to '**.5625**' and the **Number of Instances** to '**7**'.

Click in the **Features to Pattern** box, and select **0.25 Hexagon** from the flyout FeatureManager design tree.

Click the green check mark button at the top of the **Linear Pattern** PropertyManager.

In the FeatureManager design tree, right click on the sketch under **0.25 Hexagon** and pick **Hide**.

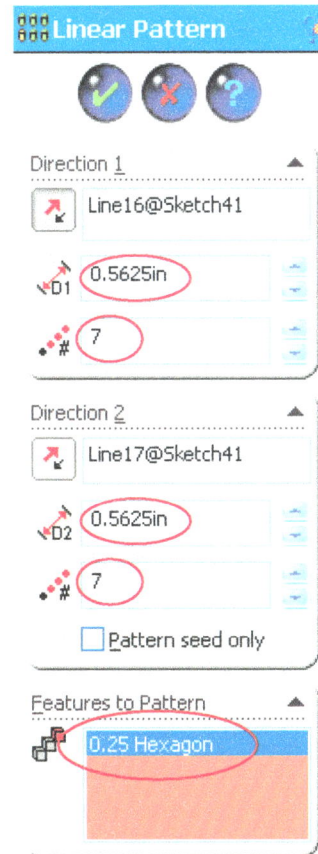

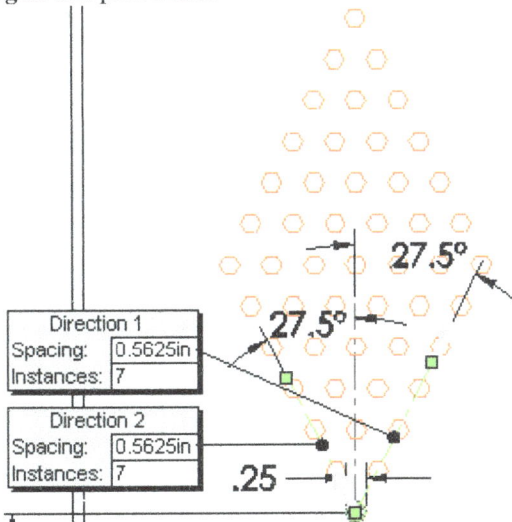

Create an Octagon

Click the **Extruded Cut** icon in the CommandManager, or pull down the "Insert" menu and pick **Cut – Extrude**.

Pick anywhere on the top surface of the part when prompted to select a plane on which to sketch the feature cross-section.

Pull down the "Tools" menu and pick **Sketch Entities – Polygon**.

In the **Polygon** PropertyManager, set the **Number of Sides** to '**8**'.

Pick on the right side of the part to indicate the starting location of the octagon. Move the cursor up and to the left to place a flat side at the bottom of the octagon and click to place the octagon.

Click the **Centerline** icon in the CommandManager, or pull down the "Tools" menu and pick **Sketch Entities – Centerline**.

Create a vertical centerline starting at the centerpoint of the octagon and going up. Then, create an angled centerline to the left starting at the centerpoint of the octagon.

Click the **Smart Dimension** icon in the CommandManager, or pull down the "Tools" menu and pick **Dimensions – Smart.**

Add a '**2.5**' horizontal dimension from the center of the octagon to the right edge of the part. Then, add '**5**' vertical dimension from the center of the octagon to the bottom of the part. Next, add a '**45**' angular dimension between the vertical centerline and the angled centerline. Then, add a '**135**' angular dimension between the vertical centerline and the top right angled side of the octagon. Finally, add a '**.375**' horizontal dimension across the width from the left side to the right side of the octagon as shown.

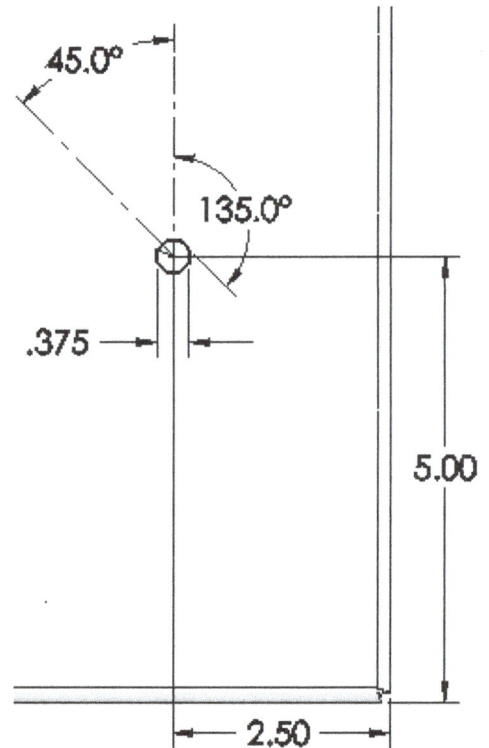

Exit the sketch by clicking the **Exit Sketch** icon in the CommandManager or in the upper right corner of the graphics area.

In the **Cut-Extrude** PropertyManager, check the **Link to thickness** check box and then click the green check mark button at the top of the **Cut-Extrude** PropertyManager.

In the FeatureManager design tree, slowly click two times (left click once, pause, left click again) on **Cut-Extrude5** to select it. Then, enter '**0.375 Octagon**' for the new name.

In the FeatureManager design tree, click on the plus to the left of **0.375 Octagon**. Right click on the sketch and pick **Show**.

A Grid of Octagons

Click the **Linear Pattern** icon in the CommandManager, or pull down the "Insert" menu and pick **Pattern/Mirror – Linear Pattern**.

In the graphics area, select the bottom half of the vertical centerline on the right to set the pattern direction for **Direction 1**.

Then, select the bottom half of the angled centerline on the left to set the pattern direction for **Direction 2**.

In the **Linear Pattern** PropertyManager, under **Direction 1**, set the **Spacing** to '**.625**' and the **Number of Instances** to '**6**'.

In the **Linear Pattern** PropertyManager, under **Direction 2**, set the **Spacing** to '**.5625**' and the **Number of Instances** to '**7**'.

Click in the **Features to Pattern** box, and select **0.375 Octagon** from the flyout FeatureManager design tree.

Click the green check mark button at the top of the **Linear Pattern** PropertyManager.

In the FeatureManager design tree, right click on the sketch under **0.375 Octagon** and pick **Hide**.

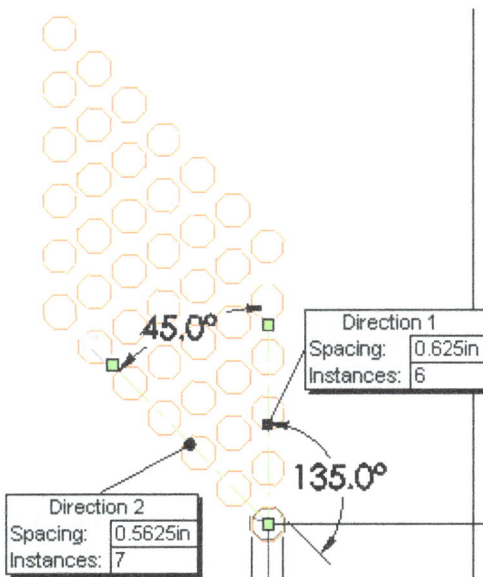

In the bottom left corner of the graphics area, change the View orientation by clicking the pull down arrow and picking **Trimetric**.

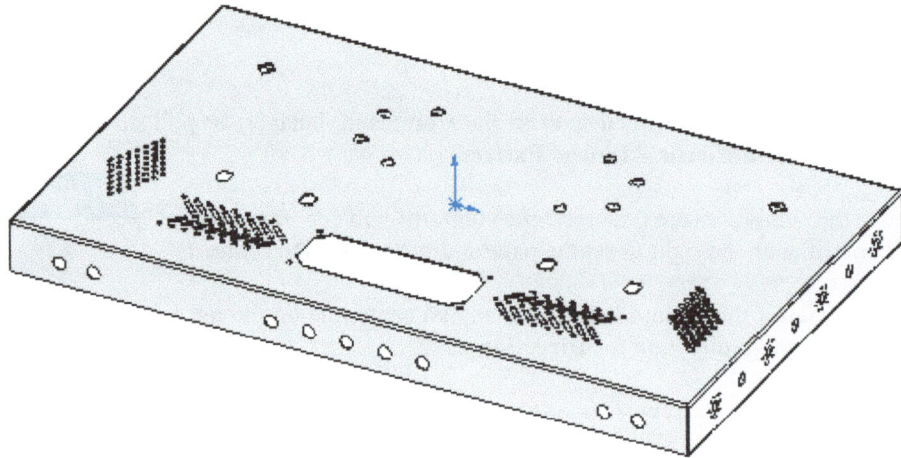

Save the Part

Click the **Save** icon in the "Standard" toolbar, or pick **Save** from the "File" pull down menu.

The **Save As** dialog box appears. In the **File name** box, type '**Front Panel**' and click **Save**.

Congratulations

You have now completed Course 2. You have seen by now that you can pattern an object in the sketch or exit the sketch and pattern the feature. The same is true with the Mirror functions. You can create a round hole with the Simple Hole command, the Cut-Extrude command, or the Hole Wizard. Yet, there are some major differences between all of these and you will want to learn which is best to use for your situation.

"SolidWorks for the Sheet Metal Guy – Course3: Unfolding" will introduce you to bend tables, material and gauge tables. Learn how to design in the flat, how to add features to the flat pattern, and fix the part for manufacturing. You will also learn how to use sheet metal parts from other CAD systems.

Index